Biochemistry of Collagens, Laminins and Elastin

Biochemistry of Collagens, Laminins and Elastin

Structure, Function and Biomarkers

Edited by

Morten A. Karsdal
Nordic Bioscience, Herlev,
Denmark and Southern Danish University,
Odense, Denmark

Co-editors

Diana J. Leeming
Kim Henriksen
Anne-Christine Bay-Jensen

AMSTERDAM • BOSTON • HEIDELBERG • LONDON
NEW YORK • OXFORD • PARIS • SAN DIEGO
SAN FRANCISCO • SINGAPORE • SYDNEY • TOKYO

Academic Press is an imprint of Elsevier

Academic Press is an imprint of Elsevier
125 London Wall, London EC2Y 5AS, United Kingdom
525 B Street, Suite 1800, San Diego, CA 92101-4495, United States
50 Hampshire Street, 5th Floor, Cambridge, MA 02139, United States
The Boulevard, Langford Lane, Kidlington, Oxford OX5 1GB, United Kingdom

Notices
Knowledge and best practice in this field are constantly changing. As new research and experience broaden our understanding, changes in research methods, professional practices, or medical treatment may become necessary.

Practitioners and researchers must always rely on their own experience and knowledge in evaluating and using any information, methods, compounds, or experiments described herein. In using such information or methods they should be mindful of their own safety and the safety of others, including parties for whom they have a professional responsibility.

To the fullest extent of the law, neither the Publisher nor the authors, contributors, or editors, assume any liability for any injury and/or damage to persons or property as a matter of products liability, negligence or otherwise, or from any use or operation of any methods, products, instructions, or ideas contained in the material herein.

Library of Congress Cataloging-in-Publication Data
A catalog record for this book is available from the Library of Congress

British Library Cataloguing-in-Publication Data
A catalogue record for this book is available from the British Library

ISBN: 978-0-12-809847-9

For information on all Academic Press publications
visit our website at https://www.elsevier.com/

Publisher: Sara Tenney
Acquisition Editor: Jill Leonard
Editorial Project Manager: Fenton Coulthurst
Production Project Manager: Edward Taylor
Designer: Christian Bilbow

Typeset by TNQ Books and Journals

Contents

List of Contributors

A. Arvanitidis Nordic Bioscience, Herlev, Denmark

C.L. Bager Nordic Bioscience, Herlev, Denmark

A.C. Bay-Jensen Nordic Bioscience, Herlev, Denmark

F. Genovese Nordic Bioscience, Herlev, Denmark

N.S. Gudmann Nordic Bioscience, Herlev, Denmark

D. Guldager Kring Rasmussen Nordic Bioscience, Herlev, Denmark

N.U.B. Hansen Nordic Bioscience, Herlev, Denmark

K. Henriksen Nordic Bioscience Biomarkers & Research, Herlev, Denmark

Y. He Nordic Bioscience, Herlev, Denmark

M.A. Karsdal Nordic Bioscience, Herlev, Denmark

S.N. Kehlet Nordic Bioscience, Herlev, Denmark

N.G. Kjeld Nordic Bioscience, Herlev, Denmark

J.H. Kristensen Nordic Bioscience, Herlev, Denmark; The Technical University of Denmark, Kongens Lyngby, Denmark

D.J. Leeming Nordic Bioscience, Herlev, Denmark

Y.Y. Luo Nordic Bioscience, Herlev, Denmark

T. Manon-Jensen Nordic Bioscience, Herlev, Denmark

J.H. Mortensen Nordic Bioscience, Herlev, Denmark

M.J. Nielsen Nordic Bioscience, Herlev, Denmark

S.H. Nielsen Nordic Bioscience, Herlev, Denmark

J.M.B. Sand Nordic Bioscience, Herlev, Denmark

A.S. Siebuhr Nordic Bioscience, Herlev, Denmark

S. Sun Nordic Bioscience, Herlev, Denmark

N. Willumsen Nordic Bioscience, Herlev, Denmark

Preface

This book on extracellular matrix (ECM) proteins is the result of appreciation and awe for matrix biology and structural proteins. These proteins are emerging as much more than passive bystanders to the fascinating life, death, and fate of cells: they control these cells.

Many researchers and their important work have been cited in this book; however, not all, and not all who deserve to be cited are included. Thus, for all who are working on collagens, laminins, and elastin, please send your references and a summary of your work to be included in future editions of this book. The aspiration of these books on the ECM is to be as complete as possible regarding research on collagen biomarkers and their biology. Please contribute to this ongoing aspiration.

The functions of many collagens still remain to be discovered and presented, both with respect to their physiological and pathophysiological roles. The hope of this book, and subsequent books, is to inspire new researchers to take the collagen challenge and present novel research and biology that are important for understanding the role of the ECM in pathological and physiological conditions.

Sincerely

Morten A. Karsdal, MSc, PhD, mBMA
Professor, University of Southern Denmark

Acknowledgments

I thank Claus Christiansen for discovering, developing, and validating (via the Food and Drug Administration) the first biomarker of the extracellular matrix (C-terminal telopeptide of type I collagen (CTX-I)). This fragment is a neo-epitope of type I collagen generated by proteolytic activity of cathepsin K, and is now recognized as the standard bone resorption marker. This discovery has inspired many researchers, including me, to discover, develop, and validate biomarkers of the ECM. Claus has always inspired us to do crazy, impossible, but focused science, with the goal of providing research that is applicable to many fields and researchers, to forward science.

I thank all past and current PhD students as well as senior researchers that have helped me with understanding and quantifying the matrix. Without your dedication and hard work in generating data and conducting assays, this book would have been impossible. Special thanks are extended to the excellent technical help in generating novel and critical assays of the matrix and for meticulous sample measurements.

This book is truly a team effort of a large group of ECM researchers, all of whom are dedicated to quantifying and understanding the matrix in both pathological and physiological conditions. Thank you all for the help with this book.

Most importantly, I thank all former and current collaborators who provided samples and engaged in discussions that helped in understanding the role of the ECM in connective tissue biology.

Lastly, I thank the Danish Research Foundation for making it possible to write this book through the support of PhD programs, research on the ECM and biomarkers, and excellence in science.

Sincerely
Morten A. Karsdal

List of Abbreviations

97-LAD	97-kDa linear IgA dermatosis antigen
aa	Amino acid
ADAM	A disintegrin and metalloproteinase
ANCA	Antineutrophil cytoplasmic antibodies
APP	Amyloid precursor protein
ASPD	Antisocial personality disorder
BACE1	β-site APP-cleaving enzyme 1
bFGF	Basic fibroblast growth factor
BM	Bethlem myopathy
BMP-1	Bone morphogenetic protein 1
BMZ	Basement membrane zone
BP180	180-kDa bullous pemphigoid antigen
BP230	230-kDa bullous pemphigoid antigen
C5M	Matrix metalloproteinase fragment of type V collagen
CCDD	Congenital cranial dysinnervation disorder
CIA	Collagen-induced arthritis
CLAC	Collagen-like amyloidogenic component
COL	Collagenous domain
COPD	Chronic obstructive pulmonary disease
DDR1	Discoidin domain receptor 1
DMD	Duchenne muscular dystrophy
ECM	Extracellular matrix
EDS	Ehlers–Danlos syndrome
eGFR	Estimated glomerular filtration rate
ELISA	Enzyme-linked immunosorbent assay
EMI	Emilin
EMID2	Emilin/multimerin domain–containing protein 2
FACIT	Fibril-associated collagens with interrupted triple helices
FN	Fibronectin type III
G	Globular
GBM	Glomerular basement membrane
HANAC	Hereditary angiopathy with nephropathy, aneurysms, and muscle cramps
HGNC	HUGO Gene Nomenclature Committee
HNE	Human neutrophil elastase
HNSCC	Squamous cell carcinoma of the head and neck
HPLC-MS	High-performance liquid chromatography–mass spectrometry
HSGAG	Heparan sulfate glycosaminoglycan
IGFBP-5	Insulin-like growth factor binding protein-5
IHC	Immunohistochemistry

IPF	Idiopathic pulmonary fibrosis
ISEMF	Intestinal subepithelial myofibroblasts
JEB	Junctional epidermolysis bullosa
KO	Knockout
LAD-1	120-kDa linear IgA dermatosis antigen
LG	Laminin globular
MI	Myocardial infarction
MIM	Mendelian inheritance in man
MMP	Matrix metalloproteinases
mRNA	Messenger ribonucleic acid
MTJ	Myotendinous junctions
NAG	N-acetyl-β-D-glucosaminidase
NC1	Noncollagenous 1
NF	Nuclear factor
NSCLC	Non-small-cell lung carcinoma
OA	Osteoarthritis
OSCC	Oral squamous cell carcinoma
P5CP	C-terminal propeptide of type V collagen
P5NP	N-terminal propeptide of type V collagen
PARP	Proline-arginine–rich protein
PDGF	Platelet-derived growth factor
Pro-C5	Neoepitope of the C-terminal propeptide of type V collagen
SNP	Single nucleotide polymorphism
SVAS	Supravalvular aortic stenosis
TACE	Tumor necrosis factor-α converting enzyme
TCR	T cell receptor
Tgase	Transglutaminase
TGF-β	Transforming growth factor-β
TSP	Thrombospondin
TSPN	Thrombospondin N-terminal-like domain
TSPN-1	N-terminal domain of thrombospondin-1
UCMD	Ullrich congenital muscular dystrophy
vWF-A	Type A domains of von Willebrand factor
α1	α1 chain
α2	α2 chain

Introduction

M.A. Karsdal[1,2]
[1]*Nordic Bioscience, Herlev, Denmark;* [2]*Southern Danish University, Odense, Denmark*

The backbone of tissues is composed of structural proteins such as collagens, laminins, and elastin. During tissue turnover, these proteins are formed and degraded in a tight equilibrium to ensure tissue health and homeostasis. Imbalances in these processes can result in fibrosis. Fibrosis can affect almost any organ or tissue. The core protein of fibrosis is collagen and other structural proteins such as laminins and elastin. Collagens are not simply structural in role: each has a unique expression pattern and some have key signaling functions in addition to their structural functions. The common denominator for collagens is the triple-helix structure, which is less pronounced in laminins. Collagens are divided into several distinct subgroups of which the fibrillar and networking collagens are the most investigated. This chapter introduces the superstructure of collagens, laminins, and elastin as well as key features of collagen biology, expression, and function.

WHY ARE COLLAGENS AND STRUCTURAL PROTEINS IMPORTANT?

Fibrosis can affect almost any organ or tissue. Fibrosis is characterized by the formation of excess connective tissue that damages the structure and function of the underlying organ or tissue and can lead to a wide variety of diseases. Fibrosis can result either from injury to tissue, in which case it manifests as scarring, or from abnormal connective tissue turnover.

Forty-five percent of all deaths in the developed world are associated with chronic fibroproliferative diseases [1,2] such as atherosclerosis and alcoholic liver disease. The common denominator of fibroproliferative diseases is dysregulated tissue remodeling, leading to the excessive and abnormal accumulation of extracellular matrix (ECM) components in affected tissues [1,3–7]. This ECM has an altered structure and signals abnormally to the cells that are embedded in it [1–5]. During fibrosis, the composition of ECM proteins and their interactions with each other and with the cells that attach to them are altered [1,7,8].

Fibrosis can affect almost any organ or tissue. Fig. 1 illustrates the major fibroproliferative diseases with a significant impact on human health [1,4,7–9].

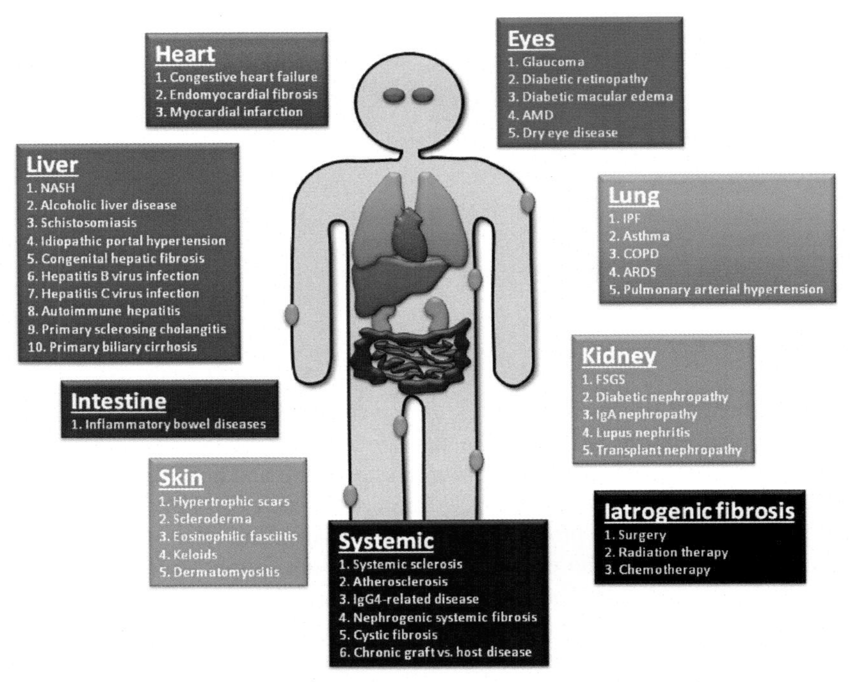

FIGURE 1 **Examples of fibroproliferative diseases in different organs.** *AMD*, age-related macular degeneration; *ARDS*, acute respiratory distress syndrome; *COPD*, chronic obstructive pulmonary disease; *IPF*, idiopathic pulmonary fibrosis; *NASH*, nonalcoholic steatohepatitis. *Reproduced with permission from Karsdal MA, Krarup H, Sand JM, Christensen PB, Gerstoft J, Leeming, DJ, et al. Review article: the efficacy of biomarkers in chronic fibroproliferative diseases – early diagnosis and prognosis, with liver fibrosis as an exemplar. Aliment Pharmacol Ther 2014;40:233–49 and Karsdal MA, Manon-Jensen T, Genovese F, Kristensen JH, Nielsen MJ, Sand, JM, et al. Novel insights into the function and dynamics of extracellular matrix in liver fibrosis. Am J Physiol Gastrointest Liver Physiol 2015.*

Fibrotic tissue was has long been considered an inactive scaffold, preventing regeneration of the affected organ. However, this perception cannot be upheld because fibrosis is neither static nor irreversible, but instead the result of a continuous remodeling that makes it susceptible to intervention [1,10,11]. The major future challenge in fibrosis will be to halt fibrogenesis and reverse advanced fibrosis without affecting tissue homeostasis or interfering with normal wound healing. Consequently, our increased understanding of the ECM, its dynamics, and the potential of fibrotic microenvironments to reverse holds promise for the development of highly specific antifibrotic therapies with minimal side effects.

Traditionally, only growth factors, cytokines, hormones, and certain other small molecules have been considered as relevant mediators of inter-, para-, and intracellular communication and signaling. However, the ECM fulfils direct and indirect paracrine or even endocrine roles. In addition to maintaining the structure of tissues, the ECM has properties that directly signal to cells. Even conceptually exclusively structural proteins such as fibrillar collagens or proteoglycans are emerging as specific signaling molecules that affect cell behavior and phenotype via cellular ECM receptors. In addition, the ECM can bind severalfold to otherwise soluble proteins, growth factors, cytokines, chemokines, or enzymes, thereby restricting or regulating their access to cells as well as specifically attracting and modulating the cells that produce these factors. Moreover, specific proteolysis can generate biologically active fragments from the ECM, while the parent molecules of the ECM are inactive. The ECM thus can control cell phenotype by functioning as a precursor bank of potent signaling fragments in addition to having a direct effect on cell phenotype through ECM–cell interactions mediated by receptors such as integrins, certain proteoglycans, or both [12–14].

The aims of this book are to (1) summarize all current data on key structural proteins of the ECM (ie, collagens, laminins, and elastin); (2) review how these molecules affect pathologies, in part, exemplified by monogenetic disorders; (3) describe selected posttranslational modifications (PTMs) of ECM proteins that result in altered signaling properties of the original ECM component; (4) discuss the novel concept that an increasing number of components of the ECM harbor cryptic signaling functions that may be viewed as endocrine functions; and (5) highlight how this knowledge can be exploited to modulate fibrotic disease.

INTRODUCTION TO THE MATRIX-INTERSTITIAL AND BASEMENT MEMBRANES

When tissue is injured, endothelial or epithelial cells on the tissue surface are destroyed, exposing the basement membrane to degradation and an influx of inflammatory cells and the deeper interstitial membrane to the risk of fibrosis (Fig. 2). The main constituents of the basement membrane are type IV collagen, laminin, and nidogen. Fragments of type IV collagen—tumstatin—have been shown to be very antiangiogenic, possibly directing recovery of the epithelium by allowing horizontal growth over the basement membrane rather than uncontrolled vertical growth into the basement and interstitial membranes. In the interstitial membrane, other collagens such as type XVIII are present. A protease-derived fragment of collagen type XVIII, endostatin [17], is the most potent natural anti-angiogenic molecule which has been show to block fibrosis in fibrotic models of the liver and lung. Other collagens such has type XV may play similar or more tissue-specific roles by releasing active protein fragments (called neoepitopes) such as restin, although this needs to be fully investigated [16].

During repair of the matrix after epithelial damage, the underlying membranes are destroyed by proteases, giving rise to new signaling molecules that may be both antifibrogenic and antiangiogenic and could potentially have other functions that are yet to be discovered.

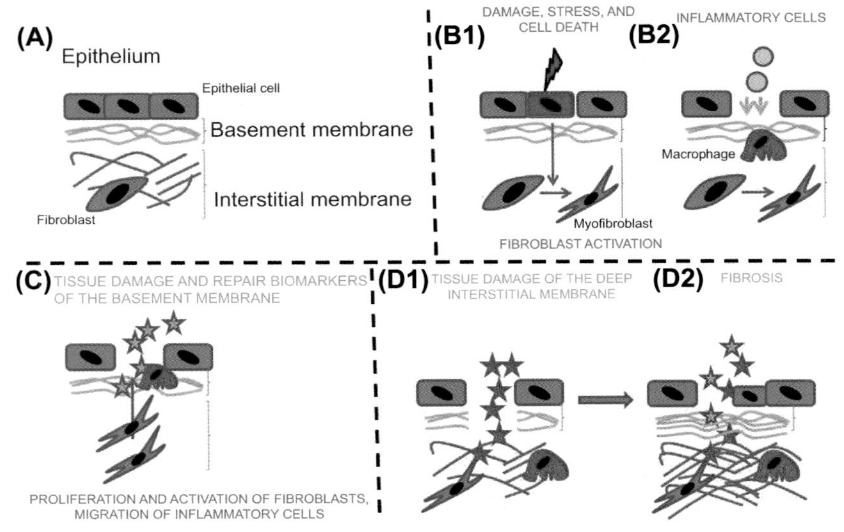

FIGURE 2 **Schematic representation of the generation of biomarkers of endothelial or epithelial cell damage from the basement and interstitial membranes.** (A) Overview of the endothelial cell layer which has well-organized basement and interstitial membranes below. (B) Cell damage results in the death of localized epithelial cells, resulting in creation of myofibroblasts from fibroblasts which proliferate and form a matrix (B1), followed by (B2) recruitment of inflammatory cells such as macrophages through the damaged epithelium. (C) Continuous cell insult results in tissue damage and denudation of the epithelium exposing the basement membrane to degradation in which fragments of the basement membrane are released. (D) Deeper tissue damage exposes the underlying interstitial membrane to degradation and its fragments are released through the basement membrane (D1). Continuous inflammation load and activation of fibroblasts results in overproduction of components of the interstitial and basement membranes in an unorganized manner, ie, in fibrosis (D2).

OVERALL STRUCTURE OF COLLAGENS

Collagens are widely expressed throughout all organs and tissues. They are the most abundant proteins in connective tissue. To date, 42 different collagen genes coding for 28 different types of collagens have been identified. Fig. 3 schematically displays the primary structure of the molecules. Collagens are trimeric molecules composed of three polypeptide α-chains which contain the repeated sequence (G–X–Y)*n*, X being frequently proline and Y hydroxyproline. These repeats allow the formation of a triple helix which is the characteristic structural feature of the collagen superfamily. Each member of the collagen family contains at least one triple-helical domain (COL) which is secreted and deposited into the extracellular matrix (ECM). Most collagens are able to form supramolecular aggregates. Besides triple-helical domains, collagens contain non-triple-helical (NC) domains, used as building blocks by other ECM proteins. The molecular structure and supramolecular assembly of collagens allow their division into major subfamilies, depending on the supramolecular structure such as depicted in Fig. 4.

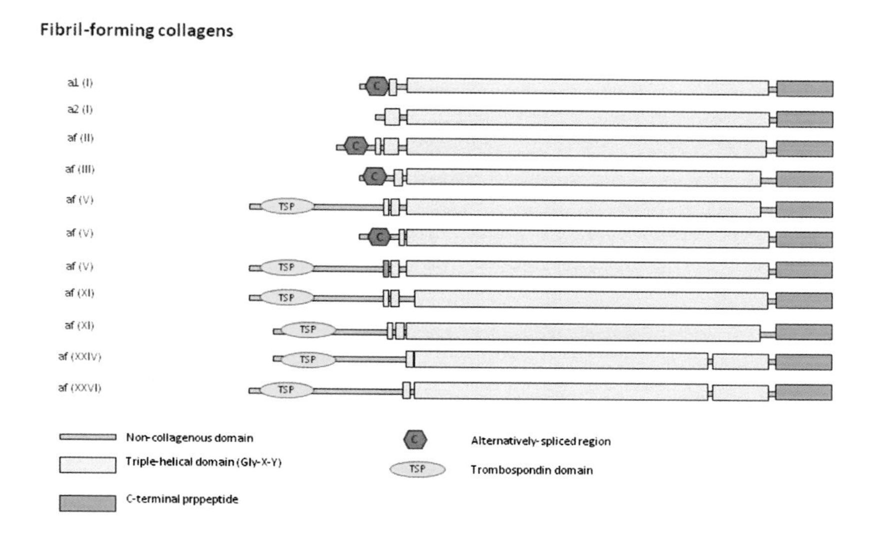

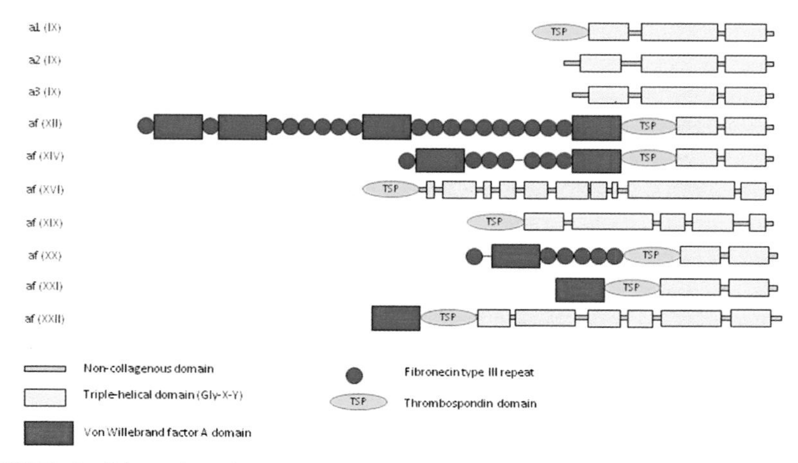

FIGURE 3 **Schematic primary structure of collagen including depiction of functional domains.** *Reproduced with permission from Ricard-Blum, S. The collagen family. Cold Spring Harb Perspect Biol 2011;3:a004978.*

Network-forming collagens

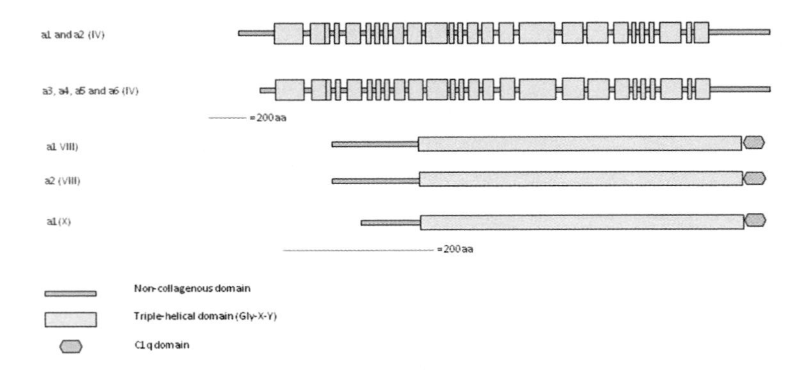

Collagens VI, VII, XXVI, and XXVIII

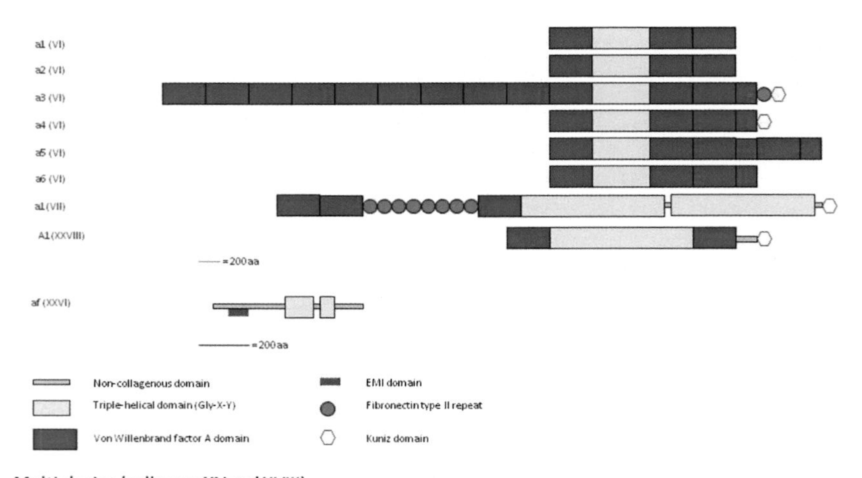

Multiplexins (collagens XV and XVIII)

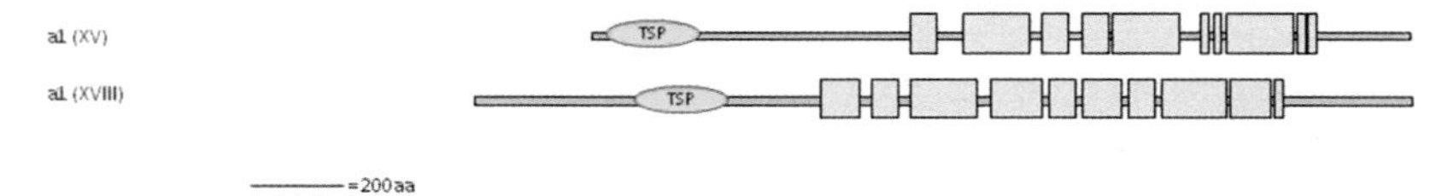

Membrane collagens

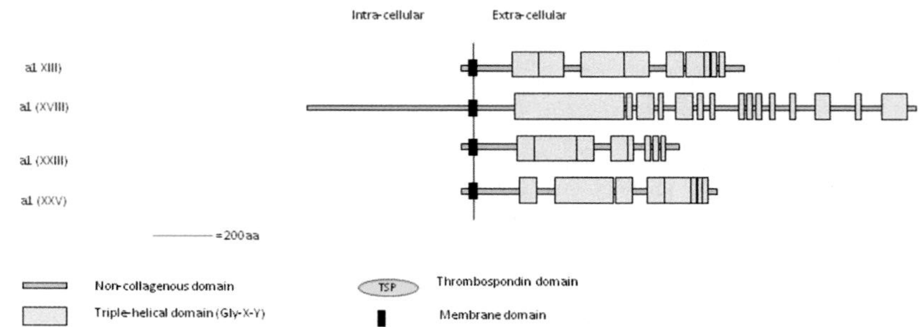

FIGURE 3 Cont'd

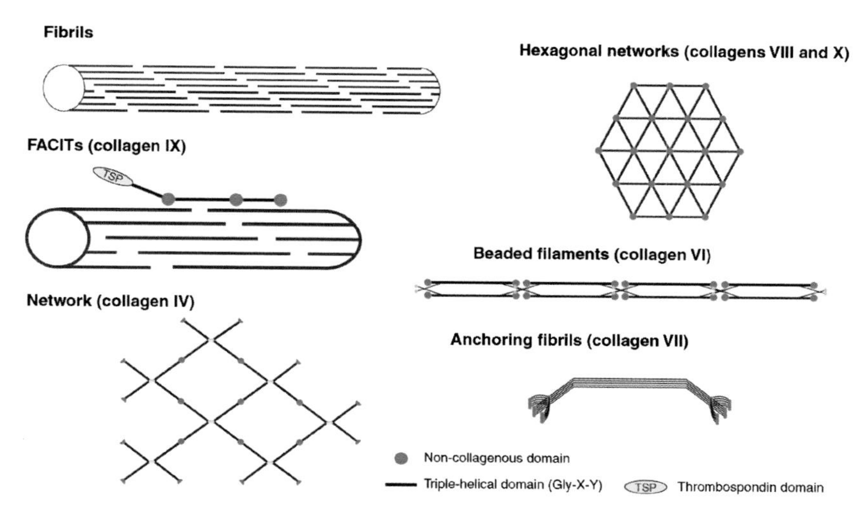

FIGURE 4 **The supermolecular structure of collagens.** *Reproduced with permission from Ricard-Blum, S. The collagen family. Cold Spring Harb Perspect Biol 2011;3:a004978.*

1. Fibril-forming collagens (I, II, III, V, XI, XXIV, XXVII);
2. Fibril-associated collagens with interrupted triple helices (FACITs) (IX, XII, XIV, XVI, XIX, XX, XXI, XXII). The FACITs do not form fibrils by themselves but they are associated with the surface of collagen fibrils.
3. Network-forming collagens (IV, VIII, X) form a pattern in which four molecules assemble via their amino-terminal 7S domain to form tetramers while two molecules assemble via their carboxy-terminal NC1 domain to form NC1 dimers
4. Membrane collagens (XIII, XVII, XXIII, XXV)

COLLAGEN SYNTHESIS AND OTHER ESSENTIALS

Collagens such as type 1 collagen are synthesized in the endoplasmic reticulum where two pro-α1 chains and one pro-α2 combine to form procollagen. This complex process $j=$is mediated by hydroxylation of prolines and lysines to stabilize the helix, secretion to the extracellular space, enzymatic removal of the N- and C-terminal propeptides, packaging the material into fibrils, and finally the formation of inter-molecular cross-links leading to the final and mechanically competent collagen fibrils. Further details on the molecular aspects of this process are outside the scope of this article, but we refer the reader to Refs. [19,20].

Extensive research has been conducted into the biosynthesis of fibril-forming collagens that are synthesized as procollagen molecules comprised of an amino-terminal propeptide followed by a short, nonhelical, N-telopeptide, a central triple helix, a C-telopeptide and a carboxy-terminal propeptide. Individual procollagen chains are subjected to numerous post-translational modifications. The heat shock protein 47 (HSP47) binds to procollagen in the endoplasmic reticulum. HSP47 is a specific molecular chaperone of procollagen [21]. The stabilization of the procollagen triple helix at body temperature requires the binding of more than 20 HSP47 molecules per triple helix [22]. It has been suggested recently that intracellular Secreted Protein Acidic and Rich in Cysteine (SPARC) might be a collagen chaperone because it binds to the triple-helical domain of procollagens and its absence leads to defects in collagen deposition in tissues [23]. Both propeptides of procollagens are cleaved during the maturation process. The N-propeptide is cleaved by procollagen N-proteinases belonging to the A Disintegrin And Metalloproteinase with Thrombospondin Motifs (ADAMTS) family, except the N-propeptide of the pro-a1(V) chain that is cleaved by the procollagen C-proteinase, which is also termed Bone Morphogenetic Protein-1 (BMP-1) [24]. BMP-1 cleaves the carboxy-terminal propeptide of procollagens except the carboxy-terminal propeptide of the pro-a1(V) chain, which is processed by furin. The telopeptides contain the sites where cross-linking occurs. This process is initiated by the oxidative deamination of lysyl and hydroxylsyl residues catalyzed by the enzymes of the lysyl oxidase family.

The top part of the figure (above the cell membrane) (Fig. 5) illustrates the intracellular events and the bottom part of the figure (below the cell membrane) illustrates the extracellular events. During the synthesis of pro-α chains in the endoplasmatic reticulum specific peptidyl lysine residues are hydroxylated to

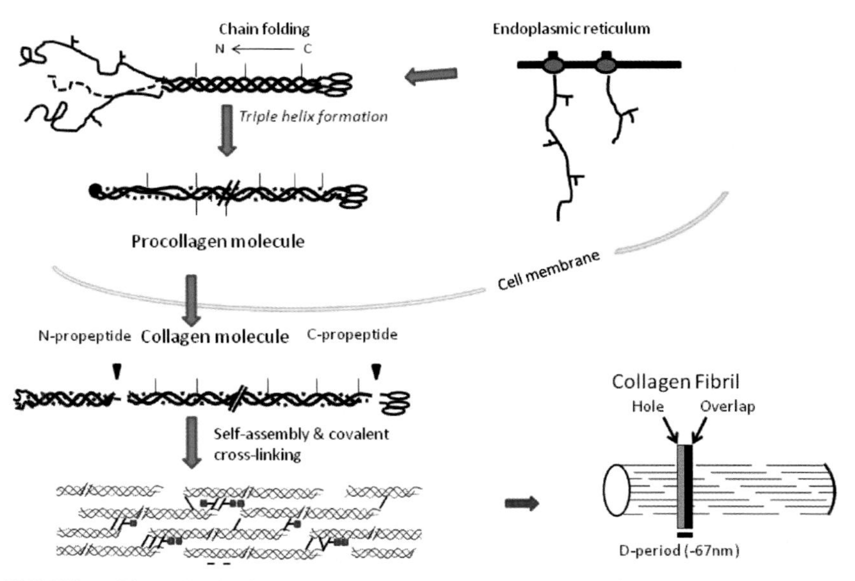

FIGURE 5 Biosynthesis of collagen. *Reproduced with permission from Yamauchi M, Sricholpech M. Lysine post-translational modifications of collagen. Essays Biochem 2012;52:113–33.*

form hydroxylysine (—OH—NH_2) and, subsequently, specific glycosylated hydroxylysine residues, this latter step being called O-linked glycosylation. For the latter, either single galactose (a red hexagon) or glucose-galactose (two red hexagons) are attached. After these and other modifications (for example hydroxylation of proline, or asparagine-linked glycosylation shown as closed circles in the C-propeptide), two pro-α1 chains (solid line) and one pro-α2 chain (dotted line) associate with one another and fold into a triple helical molecule from the C- to the N-terminus to form a procollagen molecule, packaged and secreted into the extracellular space. Then both N- and C-propeptides are cleaved to release a collagen molecule. The collagen molecules are then spontaneously self-assembled into a fibril and stabilized by covalent intra- and intermolecular covalent cross-linking. During fibrillogenesis, molecules are packed in parallel and are longitudinally staggered by an axial repeat distance, D period (~67 nm) creating two repeated regions, that is, the overlap and hole regions, in the fibril.

COLLAGEN TURNOVER AS FUNCTION OF AGE

Collagen turnover is highly affected by age, which is important for designing and interpretation of experimental settings. In fact, type I and II collagen are more than 100 fold higher in 1 months young animals as compared to 6 months old animals [26]. As illustrated in Fig. 6 (reprinted with permission from Ref. [26]), collagen turnover is drastically different in animals undergoing remodeling (rebuilding of tissues) in face of the modeling period, during building of tissues.

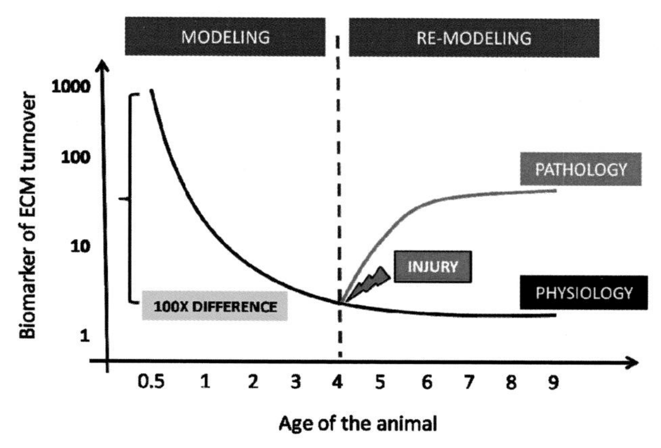

FIGURE 6 **Scheme of age-dependent ECM turnover and serum biomarker development at different age and subsequent to pathological remodeling/fibrosis.** *Reproduced with permission from Karsdal MA, Genovese F, Madsen EA, Manon-Jensen T, Schuppan D. Collagen and tissue turnover as a function of age: implications for fibrosis. J Hepatol 2015.*

This consequents in three important observations (1) the matrix composition and quality may be different in older versus younger animals (2) the relative induction of a response to an insult and pathology in older versus younger animals is higher, which is important when interpreting biomarkers as this provides better contrast as a smaller induction will not be detectable at high expression levels in young animals, albeit highly detectable in low turnover situations. This has been reported for many collage type I and II markers such as CTX-I and CTX-II [27–32]. (3) Much fibrosis research and tissue turnover research, have been conducted in younger animals which has a higher capacity for repair and turnover, which may have resulted in both false negative but also false positive observations [16,26].

Much of this regulation of these collagens is consequent to the closure of the growth plate, but also other collagens are affected 1- to 4-fold. In contrast, the interstitial type III collagen is upregulated by 1- to 2-fold [26], and basement membrane is upregulated 3-fold [26], as evident by biomarkers of type IV collagen formation and degradation. Both Type III and IV collagen stabilizes after 1–2 months after birth in rats. Other collagens such as type V and VI are not regulated [16,26,33]. Carefully designed experiments and biomarkers are lacking to provide data on the remaining of the 28 collagens.

In man, there is a strong age dependency on collagens, which however only has been carefully investigated with respect to type I and II collagen. Type I and II collagen, levels of at the age of 25, and increases after menopause, age 55 ± 5 years, consequent to the loss of sex hormones by 100–150% [32,34], which correspond to the loss in bone mineral density [31,35–39]. This is well documented in bone biology, where PINP, CTX-I have been used as biomarkers for decades [28,30,40–44].

WHY LAMININS?

Laminins as collagens are structural proteins with helical regions, albeit not a stringent as seen in the triple helical region of collagens. In collagens the helical domain consists of a strict building block which consists of three amino acids Gly-X-Y, where X and Y are often proline (Pro) and hydroxyproline (Hyp), respectively. In contrast, the triple helical structure found in laminins is made of heptads which has a less strict organization. The heptad structure represented as $(abcdefg)_n$ often has hydrophobic residues at positions a and d [45]. An example of the lower level of strictness in the heptad structure is seen when comparing the coiled-coil regions between species. Comparison of the laminin α5 chain from mammals with that of insects (that is, *drosophila melanogaster*) revealed that there was only 29% identity whereas the other domains had up to 60% identity [46]. It thus seems that the sequence motif does not rely on specific residues, but rather depends on the polarity.

Compared to collagens that are present in all compartments of the body, laminins are solely found in the basement membrane. The basement membrane is an intricate meshwork composed of laminins, collagen IV, nidogens, and sulfated proteoglycans which separates the epithelium, mesothelium and endothelium from connective tissue [47,48]. Even though the basement membrane consists of the same proteins throughout the body, different isoforms of these combine to form structurally and functionally diverse basement membranes. During the maturation of most basement membranes the composition of laminins changes. For example, as part of maturation of the glomerulus, laminin-1 is present in the early stages but is gradually replaced by laminin-10 and -11. In the final stages of maturation laminin-10 disappears, leaving laminin-11 as the sole laminin in the glomerular basement membrane (GBM) [49].

The crucial role of laminins in the basement membrane is seen during development of mice embryos where it is sufficient for the formation of basement membrane-like structures even in the absence of the other major basement membrane protein, type IV collagen [50]. Furthermore, except for the laminin α4, -β2 and -γ3 chains, deficiency of either of the laminin chains leads to early lethality [51–73]. Even for those that do not cause early lethality other major complications arise such as decreased microvessel growth and complete amyelination of nerves for Lama4-deficiency [74–77].

Laminins carry out a central role in organizing the intricate meshwork of the basement membranes. This is seen through the wide range of interaction partners which include dystroglycan [78–80], nidogens [81–83], syndecans [84–86], integrins [87–89], heparin [78,84,90], sulfatides [91] and more. The wide range of interactions combined with early lethality seen with deficiency of most laminin chains underlines the necessity of laminin presence in the basement membrane. In general, the essential role of laminins for maintenance of the basement membranes is exemplified by the lethality of most null mutations.

REFERENCES

[1] Wynn TA. Common and unique mechanisms regulate fibrosis in various fibroproliferative diseases. J Clin Invest 2007;117:524–9.

[2] Pinzani M. Welcome to fibrogenesis & tissue repair. Fibrog Tissue Repair 2008;1:1.

[3] Wynn TA, Barron L. Macrophages: master regulators of inflammation and fibrosis. Semin Liver Dis 2010;30:245–57.

[4] Wynn TA. Cellular and molecular mechanisms of fibrosis. J Pathol 2008;214:199–210.

[5] Schuppan D, Ruehl M, Somasundaram R, Hahn EG. Matrix as a modulator of hepatic fibrogenesis. Semin Liver Dis 2001;21:351–72.

[6] Schuppan D, Afdhal NH. Liver cirrhosis. Lancet 2008;371:838–51.

[7] Schuppan D, Kim YO. Evolving therapies for liver fibrosis. J Clin Invest 2013;123:1887–901.

[8] Barron L, Wynn TA. Macrophage activation governs schistosomiasis-induced inflammation and fibrosis. Eur J Immunol 2011;41:2509–14.

[9] Stone JH, Zen Y, Deshpande V. IgG4-related disease. N Engl J Med 2012;366:539–51.

[10] Troeger JS, Mederacke I, Gwak GY, Dapito DH, Mu X, Hsu CC, et al. Deactivation of hepatic stellate cells during liver fibrosis resolution in mice. Gastroenterology 2012;143:1073–83.

[11] Kisseleva T, Cong M, Paik Y, Scholten D, Jiang C, Benner C, Iwaisako K, Moore-Morris T, Scott B, Tsukamoto H, et al. Myofibroblasts revert to an inactive phenotype during regression of liver fibrosis. Proc Natl Acad Sci USA 2012;109:9448–53.

[12] Ruiz PA, Jarai G. Discoidin domain receptors regulate the migration of primary human lung fibroblasts through collagen matrices. Fibrog Tissue Repair 2012;5:3.

[13] Hynes RO. The extracellular matrix: not just pretty fibrils. Science 2009;326:1216–9.

[14] Hynes RO. Integrins: bidirectional, allosteric signaling machines. Cell 2002;110:673–87.

[15] Karsdal MA, Krarup H, Sand JM, Christensen PB, Gerstoft J, Leeming DJ, et al. Review article: the efficacy of biomarkers in chronic fibroproliferative diseases – early diagnosis and prognosis, with liver fibrosis as an exemplar. Aliment Pharmacol Ther 2014;40:233–49.

[16] Karsdal MA, Manon-Jensen T, Genovese F, Kristensen JH, Nielsen MJ, Sand JM, et al. Novel insights into the function and dynamics of extracellular matrix in liver fibrosis. Am J Physiol Gastrointest Liver Physiol 2015:G807–30.

[17] Yamaguchi Y, Takihara T, Chambers RA, Veraldi KL, Larregina AT, Feghali-Bostwick CA. A peptide derived from endostatin ameliorates organ fibrosis. Sci Transl Med 2012;4:136ra71.

[18] Ricard-Blum S. The collagen family. Cold Spring Harb Perspect Biol 2011;3:a004978.

[19] Hulmes DJS. Collagen diversity, synthesis and assembly. Springer; 2008. p. 15–47.

[20] Kivirikko KI, Myllyla R. Posttranslational enzymes in the biosynthesis of collagen: intracellular enzymes. Methods Enzymol 1982;82(Pt A):245–304.

[21] Sauk JJ, Nikitakis N, Siavash H. Hsp47 a novel collagen binding serpin chaperone, autoantigen and therapeutic target. Front Biosci 2005;10:107–18.

[22] Makareeva E, Leikin S. Procollagen triple helix assembly: an unconventional chaperone-assisted folding paradigm. PLoS One 2007;2:e1029.

[23] Martinek N, Shahab J, Sodek J, Ringuette M. Is SPARC an evolutionarily conserved collagen chaperone? J Dent Res 2007;86:296–305.

[24] Hopkins DR, Keles S, Greenspan DS. The bone morphogenetic protein 1/Tolloid-like metalloproteinases. Matrix Biol 2007;26:508–23.

[25] Yamauchi M, Sricholpech M. Lysine post-translational modifications of collagen. Essays Biochem 2012;52:113–33.

[26] Karsdal MA, Genovese F, Madsen EA, Manon-Jensen T, Schuppan D. Collagen and tissue turnover as a function of age: implications for fibrosis. J Hepatol 2015.

[27] Karsdal MA, Bay-Jensen AC, Lories RJ, Abramson S, Spector T, Pastoureau P, et al. The coupling of bone and cartilage turnover in osteoarthritis: opportunities for bone antiresorptives and anabolics as potential treatments? Ann Rheum Dis 2014;73:336–48.

[28] Henriksen K, Leeming DJ, Christiansen C, Karsdal MA. Use of bone turnover markers in clinical osteoporosis assessment in women: current issues and future options. Women's Health (Lond Engl) 2011;7:689–98.

[29] Bay-Jensen AC, Sondergaard BC, Christiansen C, Karsdal MA, Madsen SH, Qvist P. Biochemical markers of joint tissue turnover. Assay Drug Dev Technol 2010;8:118–24.

[30] Henriksen K, Bohren KM, Bay-Jensen AC, Karsdal MA. Should biochemical markers of bone turnover be considered standard practice for safety pharmacology? Biomarkers 2010;15:195–204.

[31] Henriksen K, Tanko LB, Qvist P, Delmas PD, Christiansen C, Karsdal MA. Assessment of osteoclast number and function: application in the development of new and improved treatment modalities for bone diseases. Osteoporos Int 2007;18:681–5.

[32] Karsdal MA, Henriksen K, Leeming DJ, Mitchell P, Duffin K, Barascuk N, et al. Biochemical markers and the FDA critical path: how biomarkers may contribute to the understanding of pathophysiology and provide unique and necessary tools for drug development. Biomarkers 2009;14:181–202.

[33] Jenkins RG, Simpson JK, Saini G, Bentley JH, Russell AM, Braybrooke R, et al. Longitudinal change in collagen degradation biomarkers in idiopathic pulmonary fibrosis: an analysis from the prospective, multicentre PROFILE study. Lancet Respir Med 2015;3:462–72.
[34] Iki M, Akiba T, Matsumoto T, Nishino H, Kagamimori S, Kagawa Y, et al. Reference database of biochemical markers of bone turnover for the Japanese female population. Japanese Population-based Osteoporosis (JPOS) Study. Osteoporos Int 2004;15:981–91.
[35] Karsdal MA, Qvist P, Christiansen C, Tanko LB. Optimising antiresorptive therapies in postmenopausal women: why do we need to give due consideration to the degree of suppression? Drugs 2006;66:1909–18.
[36] Karsdal MA, Tanko LB, Riis BJ, Sondergard BC, Henriksen K, Altman RD, et al. Calcitonin is involved in cartilage homeostasis: is calcitonin a treatment for OA? Osteoarthritis Cartilage 2006;14:617–24.
[37] Leeming DJ, Alexandersen P, Karsdal MA, Qvist P, Schaller S, Tanko LB. An update on biomarkers of bone turnover and their utility in biomedical research and clinical practice. Eur J Clin Pharmacol 2006;62:781–92.
[38] Oestergaard S, Sondergaard BC, Hoegh-Andersen P, Henriksen K, Qvist P, Christiansen C, et al. Effects of ovariectomy and estrogen therapy on type II collagen degradation and structural integrity of articular cartilage in rats: implications of the time of initiation. Arthritis Rheum 2006;54:2441–51.
[39] Schaller S, Henriksen K, Hoegh-Andersen P, Sondergaard BC, Sumer EU, Tanko LB, et al. In vitro, ex vivo, and in vivo methodological approaches for studying therapeutic targets of osteoporosis and degenerative joint diseases: how biomarkers can assist? Assay Drug Dev Technol 2005;3:553–80.
[40] Henriksen K, Christiansen C, Karsdal MA. Role of biochemical markers in the management of osteoporosis. Climacteric 2015;18(Suppl. 2):10–8.
[41] Henriksen K, Thudium CS, Christiansen C, Karsdal MA. Novel targets for the prevention of osteoporosis – lessons learned from studies of metabolic bone disorders. Expert Opin Ther Targets 2015;19:1575–84.
[42] Henriksen K, Karsdal MA, Martin TJ. Osteoclast-derived coupling factors in bone remodeling. Calcif Tissue Int 2014;94:88–97.
[43] Henriksen K, Christiansen C, Karsdal MA. Serological biochemical markers of surrogate efficacy and safety as a novel approach to drug repositioning. Drug Discov Today 2011;16:967–75.
[44] Karsdal MA, Byrjalsen I, Bay-Jensen AC, Henriksen K, Riis BJ, Christiansen C. Biochemical markers identify influences on bone and cartilage degradation in osteoarthritis–the effect of sex, Kellgren-Lawrence (KL) score, body mass index (BMI), oral salmon calcitonin (sCT) treatment and diurnal variation. BMC Musculoskelet Disord 2010;11:125.
[45] Beck K, Dixon TW, Engel J, Parry DA. Ionic interactions in the coiled-coil domain of laminin determine the specificity of chain assembly. J Mol Biol 1993;231:311–23.
[46] Miner JH, Lewis RM, Sanes JR. Molecular cloning of a novel laminin chain, alpha 5, and widespread expression in adult mouse tissues. J Biol Chem 1995;270:28523–6.
[47] Timpl R. Structure and biological activity of basement membrane proteins. Eur J Biochem 1989;180:487–502.
[48] Timpl R. Macromolecular organization of basement membranes. Curr Opin Cell Biol 1996;8:618–24.
[49] Miner JH, Patton BL, Lentz SI, Gilbert DJ, Snider WD, Jenkins NA, et al. The laminin alpha chains: expression, developmental transitions, and chromosomal locations of alpha1-5, identification of heterotrimeric laminins 8-11, and cloning of a novel alpha3 isoform. J Cell Biol 1997;137:685–701.

[50] Poschl E, Schlotzer-Schrehardt U, Brachvogel B, Saito K, Ninomiya Y, Mayer U. Collagen IV is essential for basement membrane stability but dispensable for initiation of its assembly during early development. Development 2004;131:1619–28.
[51] Miner JH, Li C, Mudd JL, Go G, Sutherland AE. Compositional and structural requirements for laminin and basement membranes during mouse embryo implantation and gastrulation. Development 2004;131:2247–56.
[52] Ichikawa-Tomikawa N, Ogawa J, Douet V, Xu Z, Kamikubo Y, Sakurai T, et al. Laminin alpha1 is essential for mouse cerebellar development. Matrix Biol 2012;31:17–28.
[53] Ning L, Kurihara H, de Vega S, Ichikawa-Tomikawa N, Xu Z, Nonaka R, et al. Laminin alpha1 regulates age-related mesangial cell proliferation and mesangial matrix accumulation through the TGF-beta pathway. Am J Pathol 2014;184:1683–94.
[54] Edwards MM, Mammadova-Bach E, Alpy F, Klein A, Hicks WL, Roux M, et al. Mutations in Lama1 disrupt retinal vascular development and inner limiting membrane formation. J Biol Chem 2010;285:7697–711.
[55] Miyagoe Y, Hanaoka K, Nonaka I, Hayasaka M, Nabeshima Y, Arahata K, et al. Laminin alpha2 chain-null mutant mice by targeted disruption of the Lama2 gene: a new model of merosin (laminin 2)-deficient congenital muscular dystrophy. FEBS Lett 1997;415:33–9.
[56] Kuang W, Xu H, Vachon PH, Liu L, Loechel F, Wewer UM, et al. Merosin-deficient congenital muscular dystrophy. Partial genetic correction in two mouse models. J Clin Invest 1998;102:844–52.
[57] Hager M, Gawlik K, Nystrom A, Sasaki T, Durbeej M. Laminin {alpha}1 chain corrects male infertility caused by absence of laminin {alpha}2 chain. Am J Pathol 2005;167:823–33.
[58] Yuasa K, Fukumoto S, Kamasaki Y, Yamada A, Fukumoto E, Kanaoka K, Saito K, Harada H, Arikawa-Hirasawa E, Miyagoe-Suzuki Y, et al. Laminin alpha2 is essential for odontoblast differentiation regulating dentin sialoprotein expression. J Biol Chem 2004;279:10286–92.
[59] Abrass CK, Berfield AK, Ryan MC, Carter WG, Hansen KM. Abnormal development of glomerular endothelial and mesangial cells in mice with targeted disruption of the lama3 gene. Kidney Int 2006;70:1062–71.
[60] Nguyen NM, Miner JH, Pierce RA, Senior RM. Laminin alpha 5 is required for lobar septation and visceral pleural basement membrane formation in the developing mouse lung. Dev Biol 2002;246:231–44.
[61] Miner JH, Cunningham J, Sanes JR. Roles for laminin in embryogenesis: exencephaly, syndactyly, and placentopathy in mice lacking the laminin alpha5 chain. J Cell Biol 1998;143:1713–23.
[62] Li J, Tzu J, Chen Y, Zhang YP, Nguyen NT, Gao J, Bradley M, Keene DR, Oro AE, Miner JH, et al. Laminin-10 is crucial for hair morphogenesis. EMBO J 2003;22:2400–10.
[63] Fukumoto S, Miner JH, Ida H, Fukumoto E, Yuasa K, Miyazaki H, et al. Laminin alpha5 is required for dental epithelium growth and polarity and the development of tooth bud and shape. J Biol Chem 2006;281:5008–16.
[64] Rebustini IT, Patel VN, Stewart JS, Layvey A, Georges-Labouesse E, Miner JH, et al. Laminin alpha5 is necessary for submandibular gland epithelial morphogenesis and influences FGFR expression through beta1 integrin signaling. Dev Biol 2007;308:15–29.
[65] Miner JH, Li C. Defective glomerulogenesis in the absence of laminin alpha5 demonstrates a developmental role for the kidney glomerular basement membrane. Dev Biol 2000;217:278–89.
[66] Noakes PG, Gautam M, Mudd J, Sanes JR, Merlie JP. Aberrant differentiation of neuromuscular junctions in mice lacking s-laminin/laminin beta 2. Nature 1995;374:258–62.
[67] Libby RT, Lavallee CR, Balkema GW, Brunken WJ, Hunter DD. Disruption of laminin beta2 chain production causes alterations in morphology and function in the CNS. J Neurosci 1999;19:9399–411.

[68] Noakes PG, Miner JH, Gautam M, Cunningham JM, Sanes JR, Merlie JP. The renal glomerulus of mice lacking s-laminin/laminin beta 2: nephrosis despite molecular compensation by laminin beta 1. Nat Genet 1995;10:400–6.
[69] Kuster JE, Guarnieri MH, Ault JG, Flaherty L, Swiatek PJ. IAP insertion in the murine LamB3 gene results in junctional epidermolysis bullosa. Mamm Genome 1997;8:673–81.
[70] Muhle C, Neuner A, Park J, Pacho F, Jiang Q, Waddington SN, et al. Evaluation of prenatal intra-amniotic LAMB3 gene delivery in a mouse model of Herlitz disease. Gene Ther 2006;13:1665–76.
[71] Smyth N, Vatansever HS, Murray P, Meyer M, Frie C, Paulsson M, et al. Absence of basement membranes after targeting the LAMC1 gene results in embryonic lethality due to failure of endoderm differentiation. J Cell Biol 1999;144:151–60.
[72] Meng X, Klement JF, Leperi DA, Birk DE, Sasaki T, Timpl R, et al. Targeted inactivation of murine laminin gamma2-chain gene recapitulates human junctional epidermolysis bullosa. J Invest Dermatol 2003;121:720–31.
[73] Nguyen NM, Pulkkinen L, Schlueter JA, Meneguzzi G, Uitto J, Senior RM. Lung development in laminin gamma2 deficiency: abnormal tracheal hemidesmosomes with normal branching morphogenesis and epithelial differentiation. Respir Res 2006;7:28.
[74] Thyboll J, Kortesmaa J, Cao R, Soininen R, Wang L, Iivanainen A, et al. Deletion of the laminin alpha4 chain leads to impaired microvessel maturation. Mol Cell Biol 2002;22:1194–202.
[75] Wallquist W, Plantman S, Thams S, Thyboll J, Kortesmaa J, Lannergren J, et al. Impeded interaction between Schwann cells and axons in the absence of laminin alpha4. J Neurosci 2005;25:3692–700.
[76] Yang D, Bierman J, Tarumi YS, Zhong YP, Rangwala R, Proctor TM, Miyagoe-Suzuki Y, Takeda S, Miner JH, Sherman LS, et al. Coordinate control of axon defasciculation and myelination by laminin-2 and -8. J Cell Biol 2005;168:655–66.
[77] Patton BL, Cunningham JM, Thyboll J, Kortesmaa J, Westerblad H, Edstrom L, et al. Properly formed but improperly localized synaptic specializations in the absence of laminin alpha4. Nat Neurosci 2001;4:597–604.
[78] Wizemann H, Garbe JH, Friedrich MV, Timpl R, Sasaki T, Hohenester E. Distinct requirements for heparin and alpha-dystroglycan binding revealed by structure-based mutagenesis of the laminin alpha2 LG4-LG5 domain pair. J Mol Biol 2003;332:635–42.
[79] Henry MD, Campbell KP. Dystroglycan inside and out. Curr Opin Cell Biol 1999;11:602–7.
[80] Winder SJ. The complexities of dystroglycan. Trends Biochem Sci 2001;26:118–24.
[81] Mayer U, Nischt R, Poschl E, Mann K, Fukuda K, Gerl M, et al. A single EGF-like motif of laminin is responsible for high affinity nidogen binding. EMBO J 1993;12:1879–85.
[82] Willem M, Miosge N, Halfter W, Smyth N, Jannetti I, Burghart E, et al. Specific ablation of the nidogen-binding site in the laminin gamma1 chain interferes with kidney and lung development. Development 2002;129:2711–22.
[83] Kohfeldt E, Sasaki T, Gohring W, Timpl R. Nidogen-2: a new basement membrane protein with diverse binding properties. J Mol Biol 1998;282:99–109.
[84] Utani A, Nomizu M, Matsuura H, Kato K, Kobayashi T, Takeda U, et al. A unique sequence of the laminin alpha 3 G domain binds to heparin and promotes cell adhesion through syndecan-2 and -4. J Biol Chem 2001;276:28779–88.
[85] Ogawa T, Tsubota Y, Hashimoto J, Kariya Y, Miyazaki K. The short arm of laminin gamma2 chain of laminin-5 (laminin-332) binds syndecan-1 and regulates cellular adhesion and migration by suppressing phosphorylation of integrin beta4 chain. Mol Biol Cell 2007;18:1621–33.
[86] Gersdorff N, Kohfeldt E, Sasaki T, Timpl R, Miosge N. Laminin gamma3 chain binds to nidogen and is located in murine basement membranes. J Biol Chem 2005;280:22146–53.

[87] Lian J, Dai X, Li X, He F. Identification of an active site on the laminin alpha4 chain globular domain that binds to alphavbeta3 integrin and promotes angiogenesis. Biochem Biophys Res Commun 2006;347:248–53.

[88] Kikkawa Y, Sanzen N, Fujiwara H, Sonnenberg A, Sekiguchi K. Integrin binding specificity of laminin-10/11: laminin-10/11 are recognized by alpha 3 beta 1, alpha 6 beta 1 and alpha 6 beta 4 integrins. J Cell Sci 2000;113(Pt 5):869–76.

[89] Taniguchi Y, Ido H, Sanzen N, Hayashi M, Sato-Nishiuchi R, Futaki S, et al. The C-terminal region of laminin beta chains modulates the integrin binding affinities of laminins. J Biol Chem 2009;284:7820–31.

[90] Matsuura H, Momota Y, Murata K, Matsushima H, Suzuki N, Nomizu M, et al. Localization of the laminin alpha4 chain in the skin and identification of a heparin-dependent cell adhesion site within the laminin alpha4 chain C-terminal LG4 module. J Invest Dermatol 2004;122:614–20.

[91] Taraboletti G, Rao CN, Krutzsch HC, Liotta LA, Roberts DD. Sulfatide-binding domain of the laminin A chain. J Biol Chem 1990;265:12253–8.

Chapter 1

Type I Collagen

K. Henriksen, M.A. Karsdal
Nordic Bioscience Biomarkers & Research, Herlev, Denmark

Chapter Outline

SUMMARY

Type I collagen is a fibrillar type collagen, and most likely the best investigated collagen. Type I collagen is the most abundant collagen and is the key structural composition of several tissues. It is expressed in almost all connective tissues and the predominant component of the interstitial membrane. Type I collagen mutations have documented important roles in a range of diseases, with particular focus on bone and connective tissue disease, in particular osteogenesis imperfecta and Ehlers–Danlos syndrome. Type I collagen is predominantly modified at the posttranslational level, with several crosslinks and other modifications. Several biomarkers of type I collagen have been developed, of both type I collagen degradation and formation, as surrogate makers of bone degradation and formation, respectively. Type I collagen formation is also associated with fibrosis, and fibrogenesis.

Type I collagen is the most abundant type of collagen and is expressed in almost all connective tissues. It is the major protein in bone, skin, tendon, ligament, sclera, cornea, and blood vessels. Type I collagen comprises approximately 95% of the entire collagen content of bone and about 80% of the total proteins present in bone [1].

Type I collagen is a heterotrimer molecule. In most cases it is composed of two α1 chains and one α2 chain, although an α1 homotrimer exists as a minor form. Each chain consists of more than 1000 amino acids, and the length of a collagen type I molecule is ~300nm and the width about 1–5nm. It has three domains: an N-terminal non–triple helical domain (N-telopeptide), a central triple helical domain, and a C-terminal non–triple helical domain (C-telopeptide), of which the central domain is by far the largest, comprising approximately 95% of the total molecule [2]. The triple helical domain is only possible due to the presence of glycine (G)-X-Y repeats, where X often is a proline and Y is a hydroxyproline. Glycine at every third position is essential for the correct formation of the structure [3].

Type I collagen is predominantly modified at the posttranslational level. Some of these posttranslational modifications (PTMs) are generated during the synthesis of the collagen fibrils and are essential for the mechanical competence of these fibrils [2]. Other PTMs processes, such as isomerization, racemization, enzymatic cleavage, and glycations, arise as a function of biological changes, and the end products are proving highly relevant as biomarkers of different disease aspects [3,4].

PTMs accumulating during synthesis of the collagen fibrils have been extensively reviewed [2,5]. Proline residues at the third position of the G-X-Y repeats are hydroxylated in most of the chains, whereas in some cases proline residues at the second position are also hydroxylated and these hydroxylations have proven essential for the stability of the helix [6].

A critical factor in the synthesis of a biomechanically competent type I collagen fibril is the hydroxylation of lysine residues, of which there are 38 in the α1 chain and 31 in the α2 chain [2]. The extent of hydroxylation of these residues is highly tissue dependent and ranges from 15% to 90%, with the level reflecting both tissue function and in some cases pathological changes [7]. Upon secretion of the triple helices, the N- and C-terminal peptides and propeptides are removed by enzymatic cleavage involving matrix metalloproteinases (MMPs) [8]. After cleavage, lysines and hydroxylysines in some positions, including the C terminus, are crosslinked between the individual helices, leading to the formation of extensive covalent intra- and intermolecular crosslinks, which are crucial for the mechanical competence of the collagen fibrils [9].

In addition to these PTMs, some of the hydroxylysines in type I collagen are glycosylated, although the exact extent of glycosylation is highly variable, being affected by multiple aspects such as tissue type and pathologies [10–16]. Finally, PTM processes arising spontaneously by aging, disease, or both have been described and include racemization or isomerization of aspartate and asparagine residues as well as nonenzymatic glycations, leading to the formation of advanced glycation end products (AGEs) [3]. In type I collagen, accumulation of AGEs as a function of age, elevated blood glucose levels, or both has been described. The most understood of these AGEs is pentosidine. Pentosidine is formed by condensation of glucose and the available amino groups of either arginine or lysine, which leads to the formation of a very stable crosslink between collagen chains [17]. Addition of AGE crosslinks to collagen fibers alters their mechanical properties. Studies have shown increased stiffness of tissues containing AGE-modified type I collagen fibers [18–20], hence illustrating the detrimental nature of these AGEs.

The naturally occurring PTMs caused by isomerization or racemization of aspartate or asparagine residues from α- to β-forms [21] have been studied in bone turnover. In particular, the β-form of aspartate located in the C-telopeptide has been shown to clearly reflect bone turnover [4,22]. Molecules suppressing bone turnover lead to accumulation of this PTM, whereas pathologies accelerating bone turnover reduce the level of the β-form and increase the level of the α-form [4]. Further studies show that accumulation of the β-form results in more rigid collagen structure and hence introduces stiffness in the bone. It has been speculated this stiffness is detrimental to the mechanical properties of the bone matrix [4].

In summary, type I collagen contains numerous PTMs, most of which are essential for the functionality of the fibrils. However, aging and some pathologies can introduce PTMs that are detrimental to collagens and could potentially serve as biomarkers of disease.

BIOMARKERS OF TYPE I COLLAGEN

Due to the high level of type I collagen in different tissues, particularly bone, and the well-documented turnover of type I collagen in various tissues [1], attempts have been made to measure type I collagen species in blood samples.

For 20–25 years bone mineral density (BMD) measurements and fracture rates were the tools available to detect osteoporosis and monitor treatment responses and overall bone health. However, changes in BMD are small and fractures are rare, thereby rendering assessment of new osteoporosis drugs candidates extremely slow and expensive [23].

In the mid-1990s, pioneers in the bone field working on collagen type I fragments demonstrated that these fragments provided insight into bone turnover rates, while reflecting either bone resorption or bone formation [24–26]. A truly revolutionary finding was that the presence of collagen type 1 fragments very soon after patients initiated treatment with antiresorptive drugs indicated their response to these medications, and importantly was shown to predict BMD changes over multiple years [25,27].

The implementation of the below-described bone turnover markers altered the bone field dramatically and partially as a result, osteoporosis is now a disease with multiple choices for treatment. This landmark example of how extracellular matrix turnover markers, in this case type I collagen turnover, led at least partially to additional treatments for osteoporosis, underscores the potential clinical value of yet other extracellular matrix turnover markers for different diseases.

The biomarkers related to type I collagen fall in two categories: synthesis biomarkers, such as amino-terminal propeptide of procollagen type I (PINP) and carboxy-terminal propeptide of procollagen type I (PICP); and degradation biomarkers, such as C-terminal telopeptide of type I collagen (CTX-I), type I collagen–derived crosslinked carboxy-terminal telopeptide; (ICTP), and type I collagen neoepitope (C1M) [28,29].

The collagen synthesis markers are the propeptides of type I collagen, which as described earlier are released during synthesis of the collagen molecule. Several studies have shown that these two markers reflect synthesis of bone matrix to a large extent. PINP is recommended by the International Osteoporosis Foundation (IOF) and the International Federation for Clinical Chemistry and Laboratory Medicine (IFCC) to be used in clinical studies of drugs affecting bone turnover, due to the large number of studies confirming and validating its relevance [29,30]. PINP has been shown to be elevated 100–150% during treatment with bone anabolic drugs. In these patients, bone formation was confirmed by histomorphometry. However, the increase in PINP is detectable markedly faster [30] than other bone formation parameters; hence, it is used in most trials of drugs affecting bone turnover [28,29]. An interesting study in rats indicated that PINP does not solely arise from bone; it can also be released during fibrotic changes in the liver [31]. This finding correlates well with the widespread expression pattern of type I collagen and underscores the need to know which tissues are affected by a given drug to fully understand the PINP response [28].

PICP is also used in clinical trials of bone formation drugs where its levels also rise with bone formation. However, its accuracy has not been studied to the same extent as that of PINP [29]. All these different biomarkers of collagen type 1 are visualized in Fig. 1.1.

Type I collagen degradation biomarkers have been studied for around 20 years due to their relationship to bone loss in osteoporosis [28]. Two types of type I collagen degradation markers exist: (1) cathepsin K–generated type I collagen fragments [CTX and collagen type 1 crosslinked N-telopeptide (NTX)] and (2) MMP-generated type I collagen fragments (C1M and ICTP). These markers reflect very different clinical aspects [28,32]. The cathepsin K–generated fragments reflect bone resorption and have been used extensively in clinical studies of antiresorptive drugs for osteoporosis [28]. The rationale for this resides within the highly specific combination of the enzyme cathepsin K, which is highly expressed by bone-resorbing osteoclasts, and type I collagen, which is present in massive quantities in bone [4,23,33]. Studies have carefully documented that CTX and NTX are generated directly by osteoclasts during bone resorption and that their levels reflect bone resorption in all systems from in vitro through in vivo to clinical studies (Fig. 1.2A–F) [28,34–36]. CTX has been studied extensively from applications in drug development programs in vitro, where it clearly is reduced as a function of antiresorptive compounds, and importantly where it correlates strongly with the presence of resorption pits on the actual bone slices (Fig. 1.2A–C) [34,37,38]. In addition, studies in

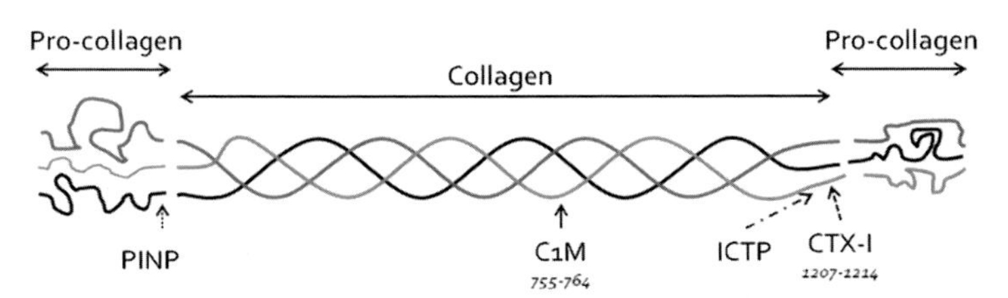

FIGURE 1.1 Schematic illustration of the localization of type I collagen neoepitopes.

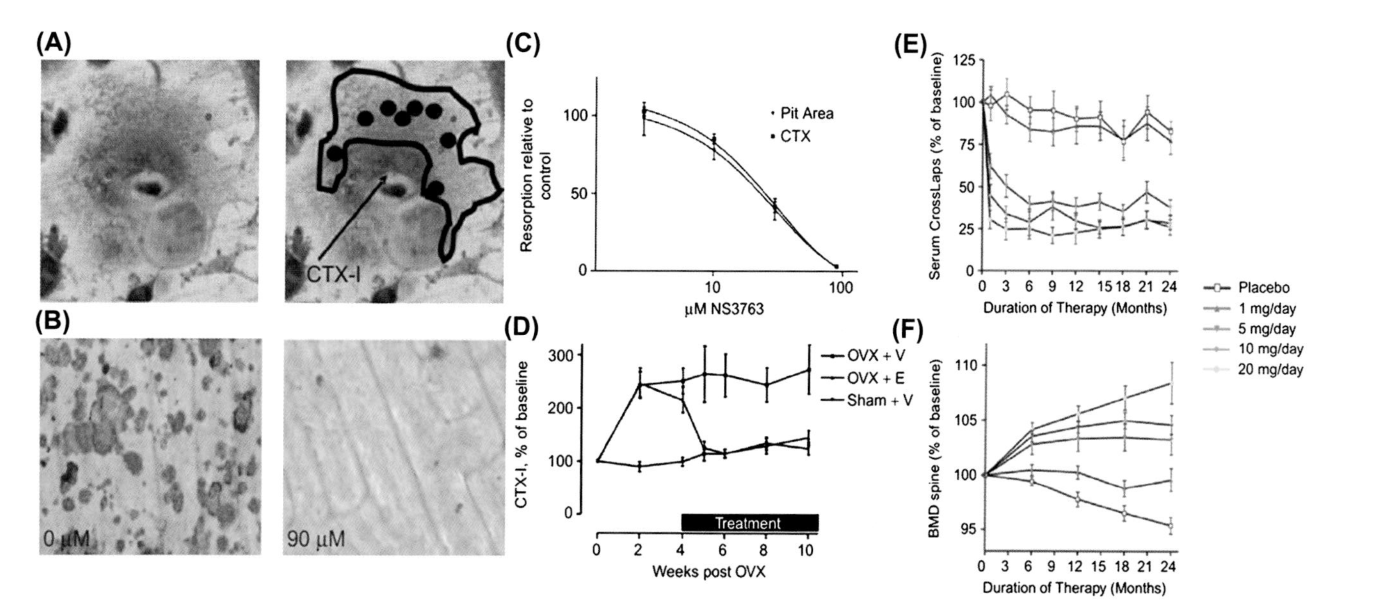

FIGURE 1.2 (A) Illustration of C-terminal telopeptide of type I collagen (CTX-I) being released from an actively resorbing osteoclast (highlighted in black outline). (B) Pictures of resorption pits taken in the presence or absence of a chloride channel inhibitor at 90 μM, clearly showing the absence of pits when resorption is blocked. (C) Plots of CTX vs. pit area scored histologically in an experiment testing a dose response of a chloride channel inhibitor. (D) CTX-I levels in serum from the aged ovariectomized rat model over time. OVX + V indicates ovariectomy with vehicle, OVX + E indicates estrogen treatment, and sham indicates the control condition. The data illustrate the OVX-induced increase in CTX-I and also show that estrogen potently reduces CTX. (E and F) Data from a clinical study of alendronate showing the dose-dependent effects on CTX-I, which are present after 1–3 months and which are predictive of bone mineral density changes over 24 months (F). *The figures are modified with permission from Karsdal MA, Henriksen K, Leeming DJ, Mitchell P, Duffin K, Barascuk N, et al. Biochemical markers and the FDA Critical Path: how biomarkers may contribute to the understanding of pathophysiology and provide unique and necessary tools for drug development. Biomarkers 2009;14:181–202; Ravn P, Clemmesen B, Christiansen C. Biochemical markers can predict the response in bone mass during alendronate treatment in early post-menopausal women. Alendronate Osteoporosis Prevention Study Group. Bone 1999;24:237–44; Ravn P, Hosking D, Thompson D, Cizza G, Wasnich RD, McClung M, et al. Monitoring of alendronate treatment and prediction of effect on bone mass by biochemical markers in the early postmenopausal intervention cohort study. J Clin Endocrinol Metab 1999;84:2363–8; Schaller S, Henriksen K, Sveigaard C, Heegaard AM, Helix N, Stahlhut M, et al. The chloride channel inhibitor n53736 prevents bone resorption in ovariectomized rats without changing bone formation. J Bone Miner Res 2004;19:1144–53.*

the ovariectomized rat model of osteoporosis have clearly shown that CTX is elevated as a function of ovariectomy, whereas it responds to inhibitors of bone resorption such as estrogen, selective estrogen receptor modulators, and alendronate (Fig. 1.2D) [38a]. In the clinical setting, one of the seminal findings is from a study by Ravn et al., who demonstrated that reductions in CTX seen after just 1 month of therapy are predictive of BMD changes occurring over future years (Fig. 1.2E–F) [25,27].

However, since the fragments are rapidly cleared from the system and since bone resorption is affected both by food intake and diurnal variation, it is essential that the fragments are measured in samples collected after fasting and in the morning [29,39,40]. Samples collected in this way have revealed that CTX-I levels, approximately 90% of which are derived directly from bone resorption, can with a high sensitivity provide information about the efficacy of antiresorptive drugs, indicate which patients will experience the fastest bone loss, and indicate risk for fractures [28]. Therefore, CTX-I is recommended as the biomarker of choice for bone resorption by the IOF and the IFCC.

Interestingly, the CTX-epitope contains an isomerization site. Two forms of the isomer, an α-form and a β-form, exist [4,22]. Studies have shown that the β-form not only accumulates with age but also as a function of agents that dramatically lower bone turnover [41]. In contrast, studies have shown that in conditions dramatically accelerating bone turnover, such as osteolytic bone metastases, the levels of α-CTX were dramatically elevated [42,43]. Data support the α/β ratio is an indicator of bone type I collagen age. It has been speculated that this ratio can serve as an index of bone quality as a function of antiresorptive therapies [4], underscoring the clinical utility of these type I collagen degradation markers.

The other collagen degradation markers are fragments generated by MMPs, and these markers are ICTP and C1M. ICTP has been studied over a long period. Although it initially was thought of as a bone resorption marker, it has now been clearly demonstrated that it does not derive directly from bone resorption, but is only generated by osteoclasts in conditions where the activity of cathepsin K is inhibited [44–46]. Under normal conditions, cathepsin K activity destroys the ICTP collagen fragment [47]. Other studies, however, have indicated that ICTP is elevated as a function of osteolytic tumors and rheumatoid arthritis [32,43].

C1M is a novel MMP-generated type I collagen fragment. It is not related to bone turnover as it is not released as a function of aggressive bone resorption as seen in osteolytic disease [48]. Studies have shown that C1M is closely related to chronic inflammation and therefore has potential as a biomarker across multiple diseases, including rheumatoid arthritis, osteoarthritis, and various forms of fibrosis [48–52]. These data clearly support that MMP-mediated destruction of type I collagen is a highly pathologically relevant process and that monitoring fragments of type I collagen provides clinical value.

A key aspect of type I collagen, found so abundantly in the body, is its post-translational modifications. These modifications are essential for correct synthesis and structural integrity of collagens, tissue-specific functionality, and application as biomarkers of different pathologies.

Type I Collagen	Description	References
Gene name and number	COL1A1, location 17q21.3-q22 COL1A2, location 7q21.3-22.1	Gene ID: 1277 Gene ID: 1278
Mutations with diseases in humans	Osteogenesis imperfecta I-IV Ehlers–Danlos Caffey disease	[53–58]
Tissue distribution in healthy states	Ubiquitous	[1]
Tissue distribution in pathological affected states	Ubiquitous	[1]
Special domains	Like other fibrillar collagens it consist of three NC domains [1–3] plus two Col domains [1,2]	[2,3]
Special neoepitopes	N- and C-terminal propeptides and N- and C-terminal degradation peptides	[2,3]
Protein structure and function	Type I collagen is a heterotrimer molecule and in most cases is composed of two α1 chains and one α2 chain, albeit an α1 homotrimer exists as a minor form. Each chain consists of more than 1000 amino acids. Glycines at every third position of the helical domain are crucial for the helix structure	[2,3]
	Essential component for the mechanical competence of the bone extracellular matrix, but also a key structural component of many other tissues. Full function not yet clear	[6,9,55,59–61]
Binding proteins	Integrins, proteoglycans, and many more	[6,9]
Known central function	Main organic component of bone, indispensable for bone integrity	[6,9]
Animals models	COL1A2-deficient mice (oim mice), collagenase-resistant collagen I mouse	[2,3]
Biomarkers	α- and β-CTX-I, NTX, ICTP, PINP, PICP, C1M	[28,29]

COL, Collagen; *CTX-1*, C-terminal telopeptide of type I collagen; *NTX*, collagen type 1 crosslinked N-telopeptide; *ICTP*, type I collagen–derived crosslinked carboxy-terminal telopeptide; *PICP*, amino-terminal propeptide of procollagen type I; *C1M*, type I collagen neoepitope.

REFERENCES

[1] Niyibizi C, Eyre DR. Structural characteristics of cross-linking sites in type V collagen of bone. Chain specificities and heterotypic links to type I collagen. Eur J Biochem 1994;224:943–50.

[2] Yamauchi M, Sricholpech M. Lysine post-translational modifications of collagen. Essays Biochem 2012;52:113–33.

[3] Viguet-Carrin S, Garnero P, Delmas PD. The role of collagen in bone strength. Osteoporos Int 2006;17:319–36.

[4] Leeming DJ, Henriksen K, Byrjalsen I, Qvist P, Madsen SH, Garnero P, et al. Is bone quality associated with collagen age? Osteoporos Int September 2009;20(9):1461–70.

[5] Kivirikko KI, Myllyla R. Posttranslational enzymes in the biosynthesis of collagen: intracellular enzymes. Methods Enzymol 1982;82(Pt A):245–304.

[6] Shoulders MD, Raines RT. Collagen structure and stability. Annu Rev Biochem 2009;78: 929–58.
[7] Uzawa K, Yeowell HN, Yamamoto K, Mochida Y, Tanzawa H, Yamauchi M. Lysine hydroxylation of collagen in a fibroblast cell culture system. Biochem Biophys Res Commun 2003;305:484–7.
[8] Hulmes DJS. Collagen diversity, synthesis and assembly. Springer; 2008. p. 15–47.
[9] Yamauchi M, Shiiba M. Lysine hydroxylation and cross-linking of collagen. Methods Mol Biol 2008;446:95–108.
[10] Moro L, Romanello M, Favia A, Lamanna MP, Lozupone E. Posttranslational modifications of bone collagen type I are related to the function of rat femoral regions. Calcif Tissue Int 2000;66:151–6.
[11] Schofield JD, Freeman IL, Jackson DS. The isolation, and amino acid and carbohydrate composition, of polymeric collagens prepared from various human tissues. Biochem J 1971;124:467–73.
[12] Toole BP, Kang AH, Trelstad RL, Gross J. Collagen heterogeneity within different growth regions of long bones of rachitic and non-rachitic chicks. Biochem J 1972;127:715–20.
[13] Michalsky M, Norris-Suarez K, Bettica P, Pecile A, Moro L. Rat cortical and trabecular bone collagen glycosylation are differently influenced by ovariectomy. Biochem Biophys Res Commun 1993;192:1281–8.
[14] Tenni R, Valli M, Rossi A, Cetta G. Possible role of overglycosylation in the type I collagen triple helical domain in the molecular pathogenesis of osteogenesis imperfecta. Am J Med Genet 1993;45:252–6.
[15] Lehmann HW, Wolf E, Roser K, Bodo M, Delling G, Muller PK. Composition and post-translational modification of individual collagen chains from osteosarcomas and osteofibrous dysplasias. J Cancer Res Clin Oncol 1995;121:413–8.
[16] Brinckmann J, Notbohm H, Tronnier M, Acil Y, Fietzek PP, Schmeller W, et al. Overhydroxylation of lysyl residues is the initial step for altered collagen cross-links and fibril architecture in fibrotic skin. J Invest Dermatol 1999;113:617–21.
[17] Sell DR, Monnier VM. Structure elucidation of a senescence cross-link from human extracellular matrix. Implication of pentoses in the aging process. J Biol Chem 1989;264:21597–602.
[18] Tomasek JJ, Meyers SW, Basinger JB, Green DT, Shew RL. Diabetic and age-related enhancement of collagen-linked fluorescence in cortical bones of rats. Life Sci 1994;55:855–61.
[19] Wang X, Shen X, Li X, Agrawal CM. Age-related changes in the collagen network and toughness of bone. Bone 2002;31:1–7.
[20] Vashishth D, Gibson GJ, Khoury JI, Schaffler MB, Kimura J, Fyhrie DP. Influence of nonenzymatic glycation on biomechanical properties of cortical bone. Bone 2001;28:195–201.
[21] Clarke S. Propensity for spontaneous succinimide formation from aspartyl and asparaginyl residues in cellular proteins. Int J Pept Protein Res 1987;30:808–21.
[22] Cloos PA, Fledelius C. Collagen fragments in urine derived from bone resorption are highly racemized and isomerized: a biological clock of protein aging with clinical potential. Biochem J 2000;345 Pt 3:473–80.
[23] Karsdal MA, Henriksen K, Leeming DJ, Mitchell P, Duffin K, Barascuk N, et al. Biochemical markers and the FDA Critical Path: how biomarkers may contribute to the understanding of pathophysiology and provide unique and necessary tools for drug development. Biomarkers 2009;14:181–202.
[24] Rosenquist C, Fledelius C, Christgau S, Pedersen BJ, Bonde M, Qvist P, et al. Serum cross-laps one step ELISA. First application of monoclonal antibodies for measurement in serum of bone-related degradation products from C-terminal telopeptides of type I collagen. Clin Chem 1998;44:2281–9.

[25] Ravn P, Clemmesen B, Christiansen C. Biochemical markers can predict the response in bone mass during alendronate treatment in early postmenopausal women. Alendronate Osteoporosis Prevention Study Group. Bone 1999;24:237–44.
[26] Garnero P, Hausherr E, Chapuy MC, Marcelli C, Grandjean H, Muller C, et al. Markers of bone resorption predict hip fracture in elderly women: the EPIDOS Prospective Study. J Bone Miner Res 1996;11:1531–8.
[27] Ravn P, Hosking D, Thompson D, Cizza G, Wasnich RD, McClung M, et al. Monitoring of alendronate treatment and prediction of effect on bone mass by biochemical markers in the early postmenopausal intervention cohort study. J Clin Endocrinol Metab 1999;84:2363–8.
[28] Henriksen K, Leeming DJ, Christiansen C, Karsdal MA. Use of bone turnover markers in clinical osteoporosis assessment in women: current issues and future options. Women's Health (Lond Engl) 2011;7:689–98.
[29] Vasikaran S, Eastell R, Bruyere O, Foldes AJ, Garnero P, Griesmacher A, et al. Markers of bone turnover for the prediction of fracture risk and monitoring of osteoporosis treatment: a need for international reference standards. Osteoporos Int 2011;22:391–420.
[30] Krege JH, Lane NE, Harris JM, Miller PD. PINP as a biological response marker during teriparatide treatment for osteoporosis. Osteoporos Int 2014;25:2159–71.
[31] Veidal SS, Vassiliadis E, Bay-Jensen AC, Tougas G, Vainer B, Karsdal MA. Procollagen type I N-terminal propeptide (PINP) is a marker for fibrogenesis in bile duct ligation-induced fibrosis in rats. Fibrogenesis Tissue Repair 2010;3:5.
[32] Fardellone P, Sejourne A, Paccou J, Goeb V. Bone remodelling markers in rheumatoid arthritis. Mediators Inflamm 2014;2014:484280.
[33] Karsdal MA, Henriksen K, Leeming DJ, Woodworth T, Vassiliadis E, Bay-Jensen AC. Novel combinations of Post-Translational Modification (PTM) neo-epitopes provide tissue-specific biochemical markers-are they the cause or the consequence of the disease? Clin Biochem 2010;43:793–804.
[34] Schaller S, Henriksen K, Sveigaard C, Heegaard AM, Helix N, Stahlhut M, et al. The chloride channel inhibitor n53736 prevents bone resorption in ovariectomized rats without changing bone formation. J Bone Miner Res 2004;19:1144–53.
[35] Karsdal MA, Byrjalsen I, Bay-Jensen AC, Henriksen K, Riis BJ, Christiansen C. Biochemical markers identify influences on bone and cartilage degradation in osteoarthritis – the effect of sex, Kellgren-Lawrence (KL) score, body mass index (BMI), oral salmon calcitonin (sCT) treatment and diurnal variation. BMC Musculoskelet Disord 2010;11:125.
[36] Karsdal MA, Byrjalsen I, Henriksen K, Riis BJ, Christiansen C. Investigations of inter- and intraindividual relationships between exposure to oral salmon calcitonin and a surrogate marker of pharmacodynamic efficacy. Eur J Clin Pharmacol 2010;66:29–37.
[37] Henriksen K, Gram J, Schaller S, Dahl BH, Dziegiel MH, Bollerslev J, et al. Characterization of osteoclasts from patients harboring a G215R mutation in ClC-7 causing autosomal dominant osteopetrosis type II. Am J Pathol 2004;164:1537–45.
[38] Sorensen MG, Henriksen K, Schaller S, Henriksen DB, Nielsen FC, Dziegiel MH, et al. Characterization of osteoclasts derived from CD14+ monocytes isolated from peripheral blood. J Bone Miner Metab 2007;25:36–45.
[38a] Schaller S, Henriksen K, Hoegh-Andersen P, Søndergaard BC, Sumer EU, Tanko LB, et al. In vitro, ex vivo, and in vivo methodological approaches for studying therapeutic targets of osteoporosis and degenerative joint diseases: how biomarkers can assist? Assay Drug Dev Technol. 2005;3(5):553–80.
[39] Bjarnason NH, Henriksen EE, Alexandersen P, Christgau S, Henriksen DB, Christiansen C. Mechanism of circadian variation in bone resorption. Bone 2002;30:307–13.

[40] Christgau S. Circadian variation in serum crosslaps concentration is reduced in fasting individuals. Clin Chem 2000;46:431.

[41] Byrjalsen I, Leeming DJ, Qvist P, Christiansen C, Karsdal MA. Bone turnover and bone collagen maturation in osteoporosis: effects of antiresorptive therapies. Osteoporos Int 2008;19:339–48.

[42] Leeming DJ, Delling G, Koizumi M, Henriksen K, Karsdal MA, Li B, et al. Alpha CTX as a biomarker of skeletal invasion of breast cancer: immunolocalization and the load dependency of urinary excretion. Cancer Epidemiol Biomarkers Prev 2006;15:1392–5.

[43] Leeming DJ, Koizumi M, Byrjalsen I, Li B, Qvist P, Tanko LB. The relative use of eight collagenous and noncollagenous markers for diagnosis of skeletal metastases in breast, prostate, or lung cancer patients. Cancer Epidemiol Biomarkers Prev 2006;15:32–8.

[44] Henriksen K, Sorensen MG, Nielsen RH, Gram J, Schaller S, Dziegiel MH, et al. Degradation of the organic phase of bone by osteoclasts: a secondary role for lysosomal acidification. J Bone Miner Res 2006;21:58–66.

[45] Nishi Y, Atley L, Eyre DE, Edelson JG, Superti-Furga A, Yasuda T, et al. Determination of bone markers in pycnodysostosis: effects of cathepsin K deficiency on bone matrix degradation. J Bone Miner Res 1999;14:1902–8.

[46] Eisman JA, Bone HG, Hosking DJ, McClung MR, Reid IR, Rizzoli R, et al. Odanacatib in the treatment of postmenopausal women with low bone mineral density: three-year continued therapy and resolution of effect. J Bone Miner Res February 2011;26(2):242–51.

[47] Sassi ML, Eriksen H, Risteli L, Niemi S, Mansell J, Gowen M, et al. Immunochemical characterization of assay for carboxyterminal telopeptide of human type I collagen: loss of antigenicity by treatment with cathepsin K. Bone 2000;26:367–73.

[48] Leeming D, He Y, Veidal S, Nguyen Q, Larsen D, Koizumi M, et al. A novel marker for assessment of liver matrix remodeling: an enzyme-linked immunosorbent assay (ELISA) detecting a MMP generated type I collagen neo-epitope (C1M). Biomarkers 2011;16:616–28.

[49] Siebuhr AS, Petersen KK, rendt-Nielsen L, Egsgaard LL, Eskehave T, Christiansen C, et al. Identification and characterisation of osteoarthritis patients with inflammation derived tissue turnover. Osteoarthritis Cartilage 2014;22:44–50.

[50] Siebuhr AS, Bay-Jensen AC, Leeming DJ, Plat A, Byrjalsen I, Christiansen C, et al. Serological identification of fast progressors of structural damage with rheumatoid arthritis. Arthritis Res Ther 2013;15:R86.

[51] Leeming DJ, Sand JM, Nielsen MJ, Genovese F, Martinez FJ, Hogaboam CM, et al. Serological investigation of the collagen degradation profile of patients with chronic obstructive pulmonary disease or idiopathic pulmonary fibrosis. Biomark Insights 2012;7:119–26.

[52] Jenkins RG, Simpson JK, Saini G, Bentley JH, Russell AM, Braybrooke R, et al. Longitudinal change in collagen degradation biomarkers in idiopathic pulmonary fibrosis: an analysis from the prospective, multicentre profile study. Lancet Respir Med June 2015;3(6):462–72.

[53] Marini JC, Forlino A, Cabral WA, Barnes AM, San Antonio JD, Milgrom S, et al. Consortium for osteogenesis imperfecta mutations in the helical domain of type I collagen: regions rich in lethal mutations align with collagen binding sites for integrins and proteoglycans. Hum Mutat 2007;28:209–21.

[54] Eyre DR, Weis MA. Bone collagen: new clues to its mineralization mechanism from recessive osteogenesis imperfecta. Calcif Tissue Int 2013;93:338–47.

[55] Cundy T. Recent advances in osteogenesis imperfecta. Calcif Tissue Int 2012;90:439–49.

[56] Gensure RC, Makitie O, Barclay C, Chan C, Depalma SR, Bastepe M, et al. A novel COL1A1 mutation in infantile cortical hyperostosis (Caffey disease) expands the spectrum of collagen-related disorders. J Clin Invest 2005;115:1250–7.

[57] Nuytinck L, Freund M, Lagae L, Pierard GE, Hermanns-Le T, De PA. Classical Ehlers–Danlos syndrome caused by a mutation in type I collagen. Am J Hum Genet 2000;66:1398–402.

[58] Malfait F, Symoens S, De BJ, Hermanns-Le T, Sakalihasan N, Lapiere CM, et al. Three arginine to cysteine substitutions in the pro-alpha (I)-collagen chain cause Ehlers–Danlos syndrome with a propensity to arterial rupture in early adulthood. Hum Mutat 2007;28:387–95.

[59] Nistala H, Makitie O, Juppner H. Caffey disease: new perspectives on old questions. Bone 2014;60:246–51.

[60] Ben AI, Glorieux FH, Rauch F. Genotype-phenotype correlations in autosomal dominant osteogenesis imperfecta. J Osteoporos 2011;2011:540178.

[61] De PA, Malfait F. The Ehlers–Danlos syndrome, a disorder with many faces. Clin Genet 2012;82:1–11.

Chapter 2

Type II Collagen

N.S. Gudmann, M.A. Karsdal
Nordic Bioscience, Herlev, Denmark

Chapter Outline

SUMMARY

Type II collagen is a fibrillar collagen, and the main component of cartilage. Type II collagen is the cartilage collagen; it constitutes 95% of the collagens and approximately 60% of dry weight. Mutations in type II collagen result in several types of chondrodysplasia, leading to premature osteoarthritis. Type II collagen is typically coassembled with collagen XI, where it is covalently crosslinked to collagen IX and interacts with small leucine-rich proteoglycans. Its stability and strength provide the tissue with integrity and resiliency to stress. Type II collagen cleavage is primarily mediated by collagenases of the matrix metalloproteinase family, resulting in well-described biomarkers such as C-terminal telopeptide of type II collagen and C2C. In addition, several formation makers have been developed. Type II collagen has important binding partners such as fibronectin and other collagens.

Type II collagen is transcribed from the gene Col2A1, and it is mainly expressed in the extracellular matrix (ECM) of articular cartilage, in the intervertebral discs, and to a lesser extent in the vitreous humor of the eyes. Type II collagen belongs to the fibril-forming collagens and is composed of three identical α1(II) chains $[\alpha 1(II)]_3$. Like other fibril-forming collagens, it consists of a right-handed triple helix of about 300 nm corresponding to about 1000 amino acids, and the triple helical (Gly-X-Y) repeat is the predominant motif [1]. This collagen molecule is about 1.5 nm in diameter [2]. Within collagen fibrils, the molecules are staggered in an N-to-C pattern. Like other fibril-forming collagens, type II collagen is synthesized in a procollagen form of which the helical domain of the α1(II) chain is the major part of the molecule and is termed Col1. As the collagen is formed, the procollagen domains of each end are cleaved off by C-proteinase

Biochemistry of Collagens, Laminins and Elastin.

and N-proteinase, respectively [3]. The C-propeptide is located at the C terminus and consists of a noncollagenous (NC)1 domain [4] comprised of three identical 35-kD chains, which are covalently bonded [5]. The N-propeptide is located at the N terminus. It can be subdivided into the three domains: NC2, Col2, and NC3 [4]. There are two splice variants of the N-propeptide of which type II collagen-derived N-terminal propeptide (PIIBNP) is usually the only variant expressed in healthy adults [6]. The other splice variant is known as type IIA procollagen amino terminal propeptide (PIIANP), and it is characterized by a prolongation of 69 amino acids in a cysteine-rich globular domain of the PIIBNP sequence. PIIANP expression is usually restricted to embryogenesis, but it is reexpressed in osteoarthritis [7]. An interesting feature of the PIIBNP fragment (but not the PIIANP variant) is that it apparently inhibits osteoclasts survival as well as bone resorption [8]. Another study suggests that PIIBNP induces cell death in tumor cells via interaction with integrins [9].

Type II collagen is typically coassembled with type XI collagen. It is covalently crosslinked to type IX collagen and interacts with small leucine-rich proteoglycans that influence collagen fibrillar architecture and function [10]. Its stability and strength provide the tissue with integrity and resiliency to stress [1]. It is the main collagen expressed in cartilage, where it constitutes 95% of all collagens [11] and accounts for 60% of the dry weight of adult cartilage [12]. Damage to the fibrillar meshwork may be a critical event in the pathology of arthritis, in part, due to low collagen turnover within the cartilage, which makes it difficult to regenerate [13].

When Col2A1 is inactivated in breeding transgenic mice, heterozygous offspring have a mild phenotype, whereas the monozygous mice die just before or shortly after birth. The cartilage of the monozygous mice is characterized by highly disorganized chondrocytes with a complete lack of extracellular fibrils discernible by electron microscopy. The mice lack both endochondrial bone and the epiphyseal growth plate in long bones. Interestingly, many other skeletal structures, such as the cranium and ribs, are normally developed and mineralized. These findings demonstrate that a well-organized cartilage matrix is required as a primary tissue for the development of some components of the vertebrate skeleton, but it is not essential for others, such as the cranial bones [12]. In animal models of autoimmunity, type II collagen has been proven to be very efficient as an immunogen in mice [14] and rats [15] in collagen-induced arthritis (CIA). CIA has been studied extensively because of its similarities to rheumatoid arthritis [14].

More than 30 different mutations have been identified in the type II collagen gene locus, all of which are heterozygous mutations mapping the triple helical domain of the molecule. The general clinical phenotype is identified as spondyloepiphyseal dysplasia. The severity ranges from very mild types of precocious osteoarthritis to the moderately severe disorder Kniest dysplasia to very severe achondrogenesis type II and hypochondrogenesis. Mutations, in general, lead to short stature as in Kniest dysplasia or to degenerative arthritis as in Stickler dysplasia as well as eye and inner ear abnormalities [16].

Type II collagen cleavage is primarily mediated by collagenases of the matrix metalloproteinase (MMP) family. Three of the known collagenases are MMP-1 (collagenase 1), MMP-8 (collagenase 2), and MMP-13 (collagenase 3), which cleave type II collagen between residues 775 (glycine) and 776 (leucine) [17]. Another cleavage site for MMP-1 is between residues 906 and 907 [18]. MMP-13 is thought to be the key enzyme involved in the excessive cleavage in osteoarthritis of interleukin (IL)-1–stimulated cartilage destruction [19]. MMP-9 (gelatinase B) is also able to cleave type II collagen and generate several cleavage fragments [20]. A well-characterized neoepitope is CB12-II, which has been found to induce MMP-13 in chondrocytes through the nuclear factor-κB pathway [21]. CB12-II was originally derived from a cyanide bromide–generated peptide CB12, which corresponds to the 195–218 residues of bovine type II collagen [22].

BIOMARKERS OF TYPE II COLLAGEN

As the main component of cartilage, evaluation of type II collagen degradation and formation has for several years been applied as a valuable tool for estimating cartilage turnover [23]. Several candidate neoepitopes have been selected for this purpose, including the COL2-3/4m antibody, which recognizes the three-quarter piece of MMP-1–cleaved triple helix of collagen II between residues 906 and 907 [18]. Several other neoepitopes have been identified and different assays have been established to measure these fragments as an estimate of cartilage degradation. One of the fragments is the C1,2C (COL2-3/4Cshort) recognizing the CGPOPQG (where O is hydroxyproline) sequence corresponding to the C terminus of the three-quarter length of the triple helical molecule. The levels of C1,2C are found to be significantly higher in α-chymotrypsin extracts of human osteoarthritic cartilage compared with extracts from nonarthritic patients. One weakness of the assay detecting this fragment was cross-reactivity toward peptides of types I and III collagen fragments [24]. The assay Col2-3/4m recognizes the sequence AOGEAGROGPOGP, a fragment of the CB12-II sequence [18]. The C2C (Col2-3/4$C_{long\ mono}$) assay recognizes the CGPPGPQG fragment generated by intrahelical cleavage of collagenases. This sequence is located at the carboxyl terminus of the three-quarter piece of the degraded α1(II) chain. The assay recognizes type II collagen cleaved by either MMP-1 or MMP-13. This cleaved collagen has been found to be significantly higher in the urine and serum of patients with rheumatoid arthritis compared with controls [25]. Another assay, type II collagen neoepitope (TIINE), uses the antibodies 9A4 and 5109 to distinguish between osteoarthritis and healthy controls [26]. Whereas 9A4 is specific for the C-terminal collagenase–generated neoepitope 776GPPGPQG794 [27], 5109 recognizes the sequence GEPGDDGPS upstream from the sequence recognized by 9A4. Both C2C and 9A4 antibodies have the disadvantage that their reactivity toward the neoepitope is affected by the hydroxylation of Pro771 of the collagen [25,28]. Levels of Pro771 hydroxylations seem to increase when type II collagen is pathologically affected [29,30]. A fragment from the N-terminal of type II collagen, GPPGPQG (which is also recognized by 9A4), is detected by the type II collagen neoepitope (CIINE) antibody when applied together with the 6G4 antibody raised against the CGEPGDDGPS in a sandwich enzyme-linked immunosorbent assay. This assay measures type II collagen fragments containing the selected neoepitope regardless of hydroxylation of Pro771 or fragment length and with minimal cross-reactivity toward types I and III collagen. CIINE has measured significantly elevated levels of collagen type II fragments in the urine of osteoarthritis patients compared with healthy controls [30]. The CartiLaps assay specifically measures cartilage degradation by quantifying C-terminal telopeptide of type II collagen (CTX-II) at the six-amino acid sequence EKGPDP, derived from the type II collagen C-terminal telopeptide. CTX-II concentrations have been found to be significantly increased in urinary samples of rheumatoid arthritis and osteoarthritis patients compared with healthy controls [31]. C2M evaluates type II collagen degradation by recognizing the RDGAAG fragment identified in the C-terminal region of the triple helical domain of type II collagen. The sequence was selected from mass spectrometry analyses of in vitro MMP-9–cleaved human cartilage [32]. This assay has been applied to differentiate between osteoarthritis and healthy controls [32] as well as in studies of rheumatoid arthritis [33] and patients with ankylosis spondylitis [34].

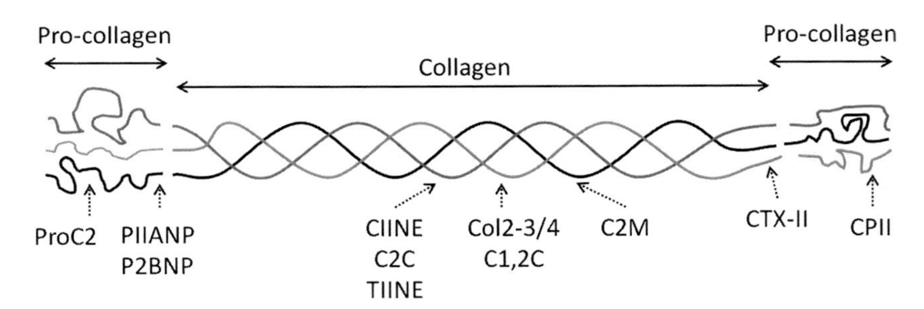

In addition to evaluating cartilage degradation, it is of interest to detect whether new cartilage is being formed; thus, the type II C-terminal propeptide (CPII) assay was developed. It evaluates levels of the C-propeptide of type II procollagen, based on rabbit antisera raised against isolated C-propeptide of bovine type II procollagen. The levels of CPII are significantly elevated in sera of rheumatoid arthritis patients and decreased in osteoarthritis compared with healthy controls [5]. Subsequently, the type IIA procollagen amino-terminal propeptide (PIIANP) assay was developed, which is sensitive to the PIIANP splice variant of the N-terminal procollagen. PIIANP is based on a polyclonal antibody in rabbits and raised against the sequence of exon 2, published of type IIA procollagen [35,36]. PIIANP is usually not expressed in healthy adults, and the assay for Pro-C2 was therefore established for evaluation of the PIIBNP splice variant of the N-terminal procollagen with the sequence QDVRQPG [37] as well as a procollagen II N-terminal propeptide (PIINP) assay that is specific for the sequence GPQGPAGEQGPRGDR and does differentiate between the two splice variants [38]. However, so far the Pro-C2 and PIINP assays have only been validated in in vitro and ex vivo models.

Type II Collagen	Description	References
Gene name and number	COL2A1, location 12q13.11	Gene ID: 1280
Mutations with diseases in humans	**In utero/congenital lethal**: • Platyspondyly Torrance type • Achondrogenesis type II • Hypochondrogenesis **Neonatal presentation**: Hypochondrogenesis (limited survival) • Spondyloepiphyseal dysplasia congenital • Kniest dysplasia • Stickler syndrome (cleft palate) • Spondyloperipheral dysplasia **Infancy or childhood**: Spondyloepimetaphyseal dysplasia (Strudwick type) • Stickler syndrome • Multiple epiphyseal dysplasia • Generalized arthritis **Adult presentation**: Avascular necrosis of the femoral head • Premature osteoarthritis	[10]
Null mutation in mice	Disorganized chondrocytes, complete lack of extracellular fibrils, no endochondrial bone or epiphyseal growth plate in long bones. Cranium and ribs were normal and mineralized.	[12]
Tissue distribution in healthy states	Nuclei pulposi and intraarticular menisci	[13,39]
Tissue distribution in pathological affected states	NA	
Special domains	Three NC domains (1–3) + two Col domains (1 and 2)	[1]
Special neoepitopes	N- and C-terminal propeptides, and N- and C-terminal	[4,5]
Protein structure and function	Homotrimer of three collagen $\alpha 1[II]_3$ chains. The helical region of each α chain is composed of 1014 amino acids, with glycine occupying every third position, where it is critical to the folding and stability of the helix. An essential component of the cartilage extracellular matrix, present in the developing cartilage anlagen and essential for endochondral bone formation	[1,10,40]

Type II Collagen	Description	References
Binding proteins	Anchorin CII; fibronectin; types IX, XI, III, and VI collagen	[2,41–44]
Known central function	Main structural protein of articular cartilage; forms the backbone of the cartilage heteropolymeric fibrils	[13]
Animals models	Collagen induced arthritis (CIA) mouse and rat models	[14,15,45]
Biomarkers	Degradation: C2M, CTX-II, TIINE, CIINE, C2C, C1,2C	[24,25,28,30–32]
	Formation: CPII, PIIANP, PIINP, Pro-C2	[5,35,37,46]

C1,C2; C2C; C2M; CIA, collagen-induced arthritis; *Col,* collagen; *CPII,* type II C-terminal propeptide; *CTX-II,* C-terminal telopeptide of type II collagen; *NA,* not applicable; *NC,* noncollagenous; *PIIANP,* type IIA procollagen amino-terminal propeptide; *PIINP,* procollagen II N-terminal propeptide; *Pro-C2; TIINE and CIINE,* type II collagen neoepitopes.

REFERENCES

[1] Gelse K, Poschl E, Aigner T. Collagens–structure, function, and biosynthesis. Adv Drug Deliv Rev 2003;55:1531–46.

[2] Mienaltowski MJ, Birk DE. Structure, physiology, and biochemistry of collagens. Adv Exp Med Biol 2014;802:5–29.

[3] Prockop DJ, Sieron AL, Li SW. Procollagen N-proteinase and procollagen C-proteinase. Two unusual metalloproteinases that are essential for procollagen processing probably have important roles in development and cell signaling. Matrix Biol 1998;16:399–408.

[4] Ricard-Blum S, Ruggiero F. The collagen superfamily: from the extracellular matrix to the cell membrane. Pathol Biol (Paris) 2005;53:430–42.

[5] Nelson F, Dahlberg L, Laverty S, Reiner A, Pidoux I, Ionescu M, et al. Evidence for altered synthesis of type II collagen in patients with osteoarthritis. J Clin Invest 1998;102: 2115–25.

[6] Sandell LJ, Morris N, Robbins JR, Goldring MB. Alternatively spliced type II procollagen mRNAs define distinct populations of cells during vertebral development: differential expression of the amino-propeptide. J Cell Biol 1991;114:1307–19.

[7] Aigner T, Zhu Y, Chansky HH, Matsen III FA, Maloney WJ, Sandell LJ. Reexpression of type IIA procollagen by adult articular chondrocytes in osteoarthritic cartilage. Arthritis Rheum 1999;42:1443–50.

[8] Hayashi S, Wang Z, Bryan J, Kobayashi C, Faccio R, Sandell LJ. The type II collagen N-propeptide, PIIBNP, inhibits cell survival and bone resorption of osteoclasts via integrin-mediated signaling. Bone 2011;49:644–52.

[9] Wang Z, Bryan J, Franz C, Havlioglu N, Sandell LJ. Type IIB procollagen NH(2)-propeptide induces death of tumor cells via interaction with integrins alpha(V)beta(3) and alpha(V)beta(5). J Biol Chem 2010;285:20806–17.

[10] Kannu P, Bateman J, Savarirayan R. Clinical phenotypes associated with type II collagen mutations. J Paediatr Child Health 2012;48:E38–43.

[11] Wu JJ, Eyre DR, Slayter HS. Type VI collagen of the intervertebral disc. Biochemical and electron-microscopic characterization of the native protein. Biochem J 1987;248:373–81.

[12] Li SW, Prockop DJ, Helminen H, Fassler R, Lapvetelainen T, Kiraly K, et al. Transgenic mice with targeted inactivation of the Col2 alpha 1 gene for collagen II develop a skeleton with membranous and periosteal bone but no endochondral bone. Genes Dev 1995;9:2821–30.

[13] Cremer MA, Rosloniec EF, Kang AH. The cartilage collagens: a review of their structure, organization, and role in the pathogenesis of experimental arthritis in animals and in human rheumatic disease. J Mol Med (Berl) 1998;76:275–88.

[14] Brand DD, Kang AH, Rosloniec EF. Immunopathogenesis of collagen arthritis. Springer Semin Immunopathol 2003;25:3–18.

[15] Trentham DE, Townes AS, Kang AH. Autoimmunity to type II collagen an experimental model of arthritis. J Exp Med 1977;146:857–68.

[16] Horton WA. Progress in human chondrodysplasias: molecular genetics. Ann N Y Acad Sci 1996;785:150–9.

[17] Dahlberg L, Billinghurst RC, Manner P, Nelson F, Webb G, Ionescu M, et al. Selective enhancement of collagenase-mediated cleavage of resident type II collagen in cultured osteoarthritic cartilage and arrest with a synthetic inhibitor that spares collagenase 1 (matrix metalloproteinase 1). Arthritis Rheum 2000;43:673–82.

[18] Hollander AP, Heathfield TF, Webber C, Iwata Y, Bourne R, Rorabeck C, et al. Increased damage to type II collagen in osteoarthritic articular cartilage detected by a new immunoassay. J Clin Invest 1994;93:1722–32.

[19] Yasuda T, Poole AR. A fibronectin fragment induces type II collagen degradation by collagenase through an interleukin-1-mediated pathway. Arthritis Rheum 2002;46:138–48.

[20] Van den Steen PE, Proost P, Grillet B, Brand DD, Kang AH, Van DJ, et al. Cleavage of denatured natural collagen type II by neutrophil gelatinase B reveals enzyme specificity, post-translational modifications in the substrate, and the formation of remnant epitopes in rheumatoid arthritis. FASEB J 2002;16:379–89.

[21] Yasuda T. Type II collagen peptide stimulates Akt leading to nuclear factor-kappaB activation: its inhibition by hyaluronan. Biomed Res 2014;35:193–9.

[22] Yasuda T, Tchetina E, Ohsawa K, Roughley PJ, Wu W, Mousa A, et al. Peptides of type II collagen can induce the cleavage of type II collagen and aggrecan in articular cartilage. Matrix Biol 2006;25:419–29.

[23] Elsaid KA, Chichester CO. Review: collagen markers in early arthritic diseases. Clin Chim Acta 2006;365:68–77.

[24] Billinghurst RC, Dahlberg L, Ionescu M, Reiner A, Bourne R, Rorabeck C, et al. Enhanced cleavage of type II collagen by collagenases in osteoarthritic articular cartilage. J Clin Invest 1997;99:1534–45.

[25] Poole AR, Ionescu M, Fitzcharles MA, Billinghurst RC. The assessment of cartilage degradation in vivo: development of an immunoassay for the measurement in body fluids of type II collagen cleaved by collagenases. J Immunol Methods 2004;294:145–53.

[26] Nemirovskiy OV, Dufield DR, Sunyer T, Aggarwal P, Welsch DJ, Mathews WR. Discovery and development of a type II collagen neoepitope (TIINE) biomarker for matrix metalloproteinase activity: from in vitro to in vivo. Anal Biochem 2007;361:93–101.

[27] Otterness IG, Downs JT, Lane C, Bliven ML, Stukenbrok H, Scampoli DN, et al. Detection of collagenase-induced damage of collagen by 9A4, a monoclonal C-terminal neoepitope antibody. Matrix Biol 1999;18:331–41.

[28] Downs JT, Lane CL, Nestor NB, McLellan TJ, Kelly MA, Karam GA, et al. Analysis of collagenase-cleavage of type II collagen using a neoepitope ELISA. J Immunol Methods 2001;247:25–34.

[29] Lippiello L, Hall D, Mankin HJ. Collagen synthesis in normal and osteoarthritic human cartilage. J Clin Invest 1977;59:593–600.

[30] Takahashi T, Naito S, Onoda J, Yamauchi A, Nakamura E, Kishino J, et al. Development of a novel immunoassay for the measurement of type II collagen neoepitope generated by collagenase cleavage. Clin Chim Acta 2012;413:1591–9.

[31] Christgau S, Garnero P, Fledelius C, Moniz C, Ensig M, Gineyts E, et al. Collagen type II C-telopeptide fragments as an index of cartilage degradation. Bone 2001;29:209–15.
[32] Bay-Jensen AC, Liu Q, Byrjalsen I, Li Y, Wang J, Pedersen C, et al. Enzyme-linked immunosorbent assay (ELISA) for metalloproteinase derived type II collagen neoepitope, CIIM–increased serum CIIM in subjects with severe radiographic osteoarthritis. Clin Biochem 2011;44:423–9.
[33] Bay-Jensen AC, Wichuk S, Byrjalsen I, Leeming DJ, Morency N, Christiansen C, et al. Circulating protein fragments of cartilage and connective tissue degradation are diagnostic and prognostic markers of rheumatoid arthritis and ankylosing spondylitis. PLoS One 2013;8:e54504.
[34] Bay-Jensen A, Leeming D, Kleyer A, Veidal S, Schett G, Karsdal M. Ankylosing spondylitis is characterized by an increased turnover of several different metalloproteinase-derived collagen species: a cross-sectional study. Rheumatol Int 2012;32:3565–72.
[35] Rousseau JC, Sandell LJ, Delmas PD, Garnero P. Development and clinical application in arthritis of a new immunoassay for serum type IIA procollagen NH_2 propeptide. Methods Mol Med 2004;101:25–37.
[36] Fukui N, McAlinden A, Zhu Y, Crouch E, Broekelmann TJ, Mecham RP, et al. Processing of type II procollagen amino propeptide by matrix metalloproteinases. J Biol Chem 2002;277:2193–201.
[37] Gudmann NS, Wang J, Hoielt S, Chen P, Siebuhr AS, He Y, et al. Cartilage turnover reflected by metabolic processing of type II collagen: a novel marker of anabolic function in chondrocytes. Int J Mol Sci 2014;15:18789–803.
[38] Madsen SH, Sondergaard BC, Bay-Jensen AC, Karsdal MA. Cartilage formation measured by a novel PIINP assay suggests that IGF-I does not stimulate but maintains cartilage formation ex vivo. Scand J Rheumatol 2009;38:222–6.
[39] Myllyharju J, Kivirikko KI. Collagens and collagen-related diseases. Ann Med 2001;33:7–21.
[40] van der Rest M, Garrone R. Collagen family of proteins. FASEB J 1991;5:2814–23.
[41] Kirsch T, Pfaffle M. Selective binding of anchorin CII (annexin V) to type II and X collagen and to chondrocalcin (C-propeptide of type II collagen). Implications for anchoring function between matrix vesicles and matrix proteins. FEBS Lett 1992;310:143–7.
[42] An B, Abbonante V, Yigit S, Balduini A, Kaplan DL, Brodsky B. Definition of the native and denatured type II collagen binding site for fibronectin using a recombinant collagen system. J Biol Chem 2014;289:4941–51.
[43] Eyre DR. The collagens of articular cartilage. Semin Arthritis Rheum 1991;21:2–11.
[44] Eyre DR, Weis MA, Wu JJ. Articular cartilage collagen: an irreplaceable framework? Eur Cell Mater 2006;12:57–63.
[45] Trentham DE, Townes AS, Kang AH, David JR. Humoral and cellular sensitivity to collagen in type II collagen-induced arthritis in rats. J Clin Invest 1978;61:89–96.
[46] Sondergaard BC, Olsen AK, Sumer EU, Qvist P, Bagger YZ, Tanko LB, et al. Calcitonin stimulates proteoglycan and collagen type II biosynthesis in articular cartilage ex vivo. Ostearthritis Cartilage 2005;13:142.

Chapter 3

Type III Collagen

M.J. Nielsen, M.A. Karsdal
Nordic Bioscience, Herlev, Denmark

Chapter Outline

SUMMARY

Type III collagen is a fibrillar collagen, and it consists of only one collagen α chain, in contrast to most other collagens. It is a homotrimer containing three α1(III) chains supercoiled around each other in a right-handed triple helix. Type III collagen is secreted by fibroblasts and other mesenchymal cell types, thus making it a major player in various inflammation-associated pathologies such as lung injury, viral and nonviral liver diseases, kidney fibrosis, hernia, and vascular disorders. Type III collagen together with type I collagen are the main constituents of the interstitial matrix. Type III collagen mutations are associated with Ehlers–Danlos syndrome, vascular deficiency, and aortic and arterial aneurysms. Several biomarkers for type III collagen formation and degradation have been developed and used extensively. In particular, for fibrosis, type III collagen formations have proven valuable.

Type III collagen is encoded by the *COL3A1* gene located on chromosome 2q31 [1]. Type III collagen belongs to the fibrillar collagen superfamily. It is a homotrimer containing three α1(III) chains supercoiled around each other in a right-handed triple helix with a characteristic 1000-residue Gly-Xaa-Yaa repeat, where Xaa and Yaa are largely represented by the amino acids proline and hydroxyproline, respectively. Glycine faces the center, whereas amino acids with longer side chains face outward, allowing close packaging along the central axis in the triple helix [2]. Type III collagen is found extensively in connective tissues such as skin, lung, liver, intestine, and the vascular system, and it is often associated with type I collagen, except in bone [3–5]. Ultrastructural analysis of tissues from mice with mutations in the *COL3A1* gene has shown that type III collagen is essential for normal type I collagen fibrillogenesis in the cardiovascular system, intestines, and skin [6].

In the extracellular matrix (ECM), type III collagen constitutes the major part of the interstitial matrix together with type I collagen. During incorporation into the ECM, N- and C-terminal propeptides are cleaved off. Cleavage of the C-terminal propeptide from soluble procollagen precursors of the fibrillar collagen is mediated by bone morphogenic protein-1 (BMP-1) and the tolloid-like proteinases, and it is the rate-limiting step in the control of fibril assembly [7]. The processing of type III collagen can be enhanced by up to 20-fold by the procollagen C-proteinase enhancers (PCPEs). Evidence suggests that PCPE1 binds exclusively to the C-terminal propeptide region of the procollagen molecule and remains bound after cleavage by BMP-1. Identification of the binding site of PCPE1 would enable the development of new strategies aimed at blocking this action and which could be a potential target for antifibrotic therapies [8].

Type III collagen is secreted by fibroblasts and other mesenchymal cell types, thus making it a major player in various inflammation-associated pathologies such as lung injury, viral and nonviral liver diseases, kidney fibrosis, hernia, and vascular disorders [9–13]. Scar tissue contains types I and III collagen with different levels of hydroxylation of lysine and glycosylation of hydroxylysine. During the process of wound healing the fibrillar collagens, including type III collagen, act as a scaffold for fibroblast attachment. This scaffolding changes their composition, ultimately leading to increased scar strength over time. Early granulation tissue contains mainly type III collagen and only a small amount of type I collagen. As wound healing continues, this ratio is altered, leading to a type I/III ratio of 1:2, and the change in immature type III collagen may result in loss of tensile strength. This shift is observed in many conditions such as liver cirrhosis, keloids, and hypertrophic scars due to an increased expression of type III procollagen mRNA [14].

Fibrillar procollagen C propeptide domains are associated with several genetic disorders of connective tissues. Mutations in this domain are particularly important as they are responsible for directing the assembly of the collagen molecule. In general, these mutations either prevent trimerization completely, leading to haploinsufficiency of the affected collagen type in heterozygotes, or they lead to abnormal procollagen assembly, which involves both the wild type and the mutant chain. One disease associated with mutations in this domain is Ehlers–Danlos syndrome (EDS) type IV, which leads to vascular deficiency [15] and aortic and arterial aneurysms. In this disease, mutations of genes may include deletion, missense mutations, and missplicing. The mutations cause disrupted formation or stability of the helix region, leading to misfolded helices. This in turn causes accumulation of defective trimers in the ECM [16]. Mouse models of EDS type IV produced by targeted deletion of *COL3A1* have been of limited use as only 10% of homozygous animals survive to adulthood, whereas heterozygous animals die from arterial rupture [6]. A recent study identified a spontaneous 185-kDa deletion in the promoter region and exons 1–39 of the murine *COL3A1* locus. This $+/COL3A1^{\Delta}$ mouse model mimics the vascular aspects of human EDS type IV with aortic dissections, which may result from aberrant collagen fibrillogenesis within the aortic wall [17].

Type III collagen serves as a ligand for several proteins, including G protein–coupled receptor-56 (GPR56), discoidin domain receptors (DDRs) 1 and 2, von Willebrand factor (vWF), and integrin α2β1. Binding to vWF and integrin α2β1 plays an important function in wound healing processes where platelets adhere to the injured vessel wall. Using collagen binding assays in vitro, researchers have identified the RGQOGVMGF sequence (O is hydroxyproline) as the major high-affinity vWF-binding site on type III collagen and it is only found in one other protein, type II collagen. The sequence is 100% conserved in mouse, rat, and cow. However, in human type I collagen, a closely related sequence is found on the α1 chain, differing by a single O-to-A substitution in position 4, suggesting that vWF may bind types I, II, and III collagen in an identical manner [18].

Interaction between GPR56 and type III collagen is essential in regulating cortical development. The interaction occurs in the developing brain, resulting in inhibition of neural migration due to a regulation of the pial basement membrane integrity and cortical lamination [19]. Type III collagen serves as a ligand for both DDR1 and DDR2 at the GVMGFO motif [20]. Normal interaction controls cell proliferation, adhesion, and migration as well as ECM remodeling, whereas aberrant interaction is associated with various diseases such as fibrosis, atherosclerosis, several types of cancers, and arthritic disorders [21].

BIOMARKERS OF TYPE III COLLAGEN

The N-terminal propeptide of type III collagen (PIIINP) is the best-known biomarker of type III collagen synthesis. It has been used as a marker of ongoing repair processes independent of etiology [22–24]. In liver fibrosis, PIIINP has shown to have acceptable diagnostic accuracy for liver fibrosis in patients with various chronic liver diseases such as alcoholic liver disease and chronic hepatitis C infection [25,26]. PIIINP has further been evaluated as a prognostic marker of liver fibrosis in limited studies. The largest prospective study of chronic hepatitis C progression is the HALT-C (Hepatitis C Antiviral Long-term Treatment against Cirrhosis) trial in which baseline levels of PIIINP were independently associated with disease progression [27].

In kidney fibrosis, two independent studies found PIIINP to be highly correlated with the extent of interstitial fibrosis [28,29]. Even though PIIINP is mainly cleared in the liver and to a lesser extent in the kidney, the urinary PIIINP marker is believed to reflect intrarenal synthesis of type III collagen since urinary PIIINP does not correlate with proteinuria [29].

Other markers of type III collagen include type III collagen neoepitope (C3M) and N-terminal propeptide of type III collagen (Pro-C3), which are novel markers based on protein fingerprint technology. The markers are designed to detect a specific cleavage fragment generated by disease-specific proteases, such as matrix metalloproteinase (MMP)-9 for C3M [30] and N-protease for

Pro-C3 [31]. Combining a protein with a protease in the assay ensures that the marker only detects the neoepitopes generated from the cleavage and not the intact protein. This means that measurements of subpools of the same protein can provide different information, emphasizing the importance of distinguishing one subpool from another [32].

Given the example of type III collagen, the PIIINP epitope of the N-terminal propeptide can be a marker of both formation and degradation since removal of the propeptide sometimes is incomplete leaving the propeptide attached to the molecule after it has been incorporated in the ECM [33]. However, targeting the N-protease cleavage site of the N-terminal propeptide, the Pro-C3 marker measures true formation of type III collagen [31], whereas targeting the MMP-9–mediated fragment, the C3M marker measures degradation of type III collagen [30].

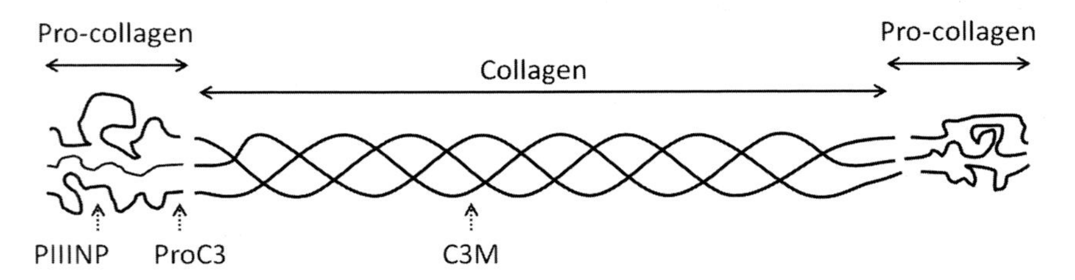

The novel two collagen markers have proven useful as diagnostic, prognostic, and efficacy of intervention tools in various diseases. In chronic kidney disease the Pro-C3 and C3M markers can be used as noninvasive diagnostic and prognostic tools to monitor ECM remodeling in both experimental and human kidney fibrosis as the two markers have been shown to be elevated in advanced stages of chronic kidney disease [34,35].

In chronic liver disease, Pro-C3 and C3M have been able to detect the degree of fibrosis in patients with various underlying etiologies of fibrosis, including hepatitis C, alcoholic liver disease, portal hypertension, and human immunodeficiency virus [36–38]. In one study the markers have even been shown to provide different clinical information related to diagnosis and prognosis in patients with chronic hepatitis C infection. When stratified according to the fibrosis score, baseline serum levels of the two markers had similar diagnostic performances as three common biomarkers of liver fibrosis: aspartate aminotransferase, alanine aminotransferase, and Fibrotest. However, when stratifying the baseline marker levels to the changes in fibrosis scores after 52 weeks, Pro-C3 was the only marker which significantly differentiated progressors from stable patients. Furthermore, patients with high Pro-C3 levels at baseline had >4 times higher odds of being progressors of fibrosis than those with low baseline levels [39].

In bile duct–ligated rats, treatment with statins was associated with a reduction in Pro-C3 levels in severe fibrosis stages [40]. In addition, in an ex vivo rat model using fibrotic precision-cut liver slices, the level of C3M was elevated in the supernatant from fibrotic liver culture compared with the control liver culture. When adding a phosphodiesterase inhibitor, the level of C3M in the fibrotic liver culture was reduced to the same level as the control liver culture [41]. These experiments suggest that the markers may serve as a read-out for efficacy of interventions to reflect restoration of the collagen synthesis from increased levels during fibrosis to a balanced healthy turnover.

Pathological type III collagen turnover has also been investigated in two types of lung diseases, idiopathic pulmonary fibrosis (IPF) and chronic obstructive pulmonary disease (COPD), by using Pro-C3 and C3M biomarkers. The first study included patients with mild COPD or IPF and healthy controls. Serum levels of C3M were significantly increased in mild COPD and IPF compared with the healthy control group [42]. This finding was confirmed in two larger studies [43,44]. Findings of the PROFILE (Prospective Study of Fibrosis In the Lung Endpoints) study, which is the largest prospective study investigating progression of IPF, showed that C3M was elevated in patients compared with healthy controls in both the discovery cohort and the validation cohort. C3M was also able to differentiate patients with progressive IPF from stable patients at 6 months and high baseline levels were further associated with overall mortality [43]. A study including patients with COPD exacerbations showed that during exacerbations the serum level of C3M was elevated compared with the level at follow-up 4 weeks later. Although no difference was observed for Pro-C3 during exacerbation compared with follow-up, the degradation/formation ratio was significantly elevated during exacerbation, suggesting an overall type III collagen destruction at these episodes [44].

Along with types I and II collagen, type III collagen is a main constituent of the joint. Increased levels of type III collagen fragments have provided novel insights into the connective tissue balance and possible pain mechanisms in rheumatoid arthritis (RA) and osteoarthritis (OA). The strong correlation between C3M and the MMP-generated fragment of C-reactive protein (CRP) observed in OA patients suggests that the increased tissue turnover is a result of local inflammatory processes in the synovial membrane of the joint [45,46]. Interestingly, in the Tocilizumab Safety and the Prevention of Structural Joint Damage (LITHE) biomarker study including more than 700 patients with moderately to severely active RA, C3M serum levels were reduced by ~70% in patients treated with 8 mg/kg tocilizumab after 52 weeks. Furthermore, a significant difference in circulating C3M levels was observed between early nonresponders, nonresponders, and responders, suggesting that specific responder profiles exist [47].

Together, these studies show that regardless of whether the disease is associated with inflammation or fibrosis, the serum concentrations of protein fragments generated by specific proteases are increased in patients compared with healthy individuals. Furthermore, elevated levels of either one of the markers or both were associated with disease progression. This suggests that the pathological ECM remodeling can be monitored in different stages, from early disease onset with inflammation to the more advanced disease stages where fibrosis is present and that ECM remodeling is a common denominator in many diseases.

Type III Collagen	Description	Reference
Gene name and number	COL3A1, chromosome 2q31	[1]
Mutations with diseases in humans	Single-base mutations cause amino acid substitutions, RNA splicing mutations, and partial gene deletions, causing vascular EDS (moderate-to-severe form). Homozygosity for the null allele causes recessive EDS.	[15,48,49]
Null mutation in mice	N.A.	
Tissue distribution in healthy states	Skin, vessel walls, lung, liver, and spleen, among others, along with type I collagen (except in bone)	[3–5]
Tissue distribution in pathologically affected states	Connective tissue fiber hyperplasia, upregulated in fibrotic tissue (lung, liver, kidney, vascular system)	[9–12,50,51]
Special domains	vWF type C domain, fibrillar collagen C-terminal domain	[52]
Special neoepitopes	N- and C-terminal propeptides	[53]
Protein structure and function	Homotrimer consisting of three α1(III) chains flanked by N- and C-terminal propeptides	[3]
Binding proteins	GPCR56, PCPE-1, DDR1, DDR2, vWF, integrin α2β1, LAIR-1, PDGF	[8,19,20,54,55]
Known central function	Cell–matrix adhesion, collagen fibril organization, ECM organization, response to TGF-β1	[18,56–58]
Animals models with protein affected	+/COl3A1 knockout	[17]
Biomarkers	PIIINP, Pro-C3, C3M	[31,59,60]

EDS, Ehlers–Danlos syndrome; *N.A.*, not applicable; *vWF*, von Willebrand factor; *CPCR*, G protein–coupled receptor; *PCPE*, procollagen C-proteinase enhancer; *DDR*, discoidin domain receptor; *LAIR*, leukocyte-associated immunoglobulin-like receptor; *PDGF*, platelet-derived growth factor; *ECM*, extracellular matrix; *TGF*, transforming growth factor; *PIIINP*, N-terminal propeptide of type III collagen; *Pro-C3*, N-terminal propeptide of type III collagen; *C3M*, type III collagen neoepitope.

REFERENCES

[1] Abayazeed A, Hayman E, Moghadamfalahi M, Cain D. Vascular type Ehlers–Danlos Syndrome with fatal spontaneous rupture of a right common iliac artery dissection: case report and review of literature. J Radiol Case Rep 2014;8:63–9.

[2] Exposito JY, Valcourt U, Cluzel C, Lethias C. The fibrillar collagen family. Int J Mol Sci 2010;11:407–26.

[3] Gelse K, Poschl E, Aigner T. Collagens–structure, function, and biosynthesis. Adv Drug Deliv Rev 2003;55:1531–46.

[4] Bao X, Zeng Y, Wei S, Wang G, Liu C, Sun Y, et al. Developmental changes of Col3a1 mRNA expression in muscle and their association with intramuscular collagen in pigs. J Genet Genomics 2007;34:223–8.

[5] Jensen LT, Host NB. Collagen: scaffold for repair or execution. Cardiovasc Res 1997; 33:535–9.

[6] Liu X, Wu H, Byrne M, Krane S, Jaenisch R. Type III collagen is crucial for collagen I fibrillogenesis and for normal cardiovascular development. Proc Natl Acad Sci USA 1997;94:1852–6.

[7] Canty EG, Kadler KE. Procollagen trafficking, processing and fibrillogenesis. J Cell Sci 2005;118:1341–53.

[8] Vadon-Le GS, Kronenberg D, Bourhis JM, Bijakowski C, Raynal N, Ruggiero F, et al. Procollagen C-proteinase enhancer stimulates procollagen processing by binding to the C-propeptide region only. J Biol Chem 2011;286:38932–8.

[9] Meduri GU, Tolley EA, Chinn A, Stentz F, Postlethwaite A. Procollagen types I and III aminoterminal propeptide levels during acute respiratory distress syndrome and in response to methylprednisolone treatment. Am J Respir Crit Care Med 1998;158:1432–41.

[10] Teare JP, Sherman D, Greenfield SM, Simpson J, Bray G, Catterall AP, et al. Comparison of serum procollagen III peptide concentrations and PGA index for assessment of hepatic fibrosis. Lancet 1993;342:895–8.

[11] Scheja A, Akesson A, Horslev-Petersen K. Serum levels of aminoterminal type III procollagen peptide and hyaluronan predict mortality in systemic sclerosis. Scand J Rheumatol 1992;21:5–9.

[12] Lin YH, Ho YL, Wang TD, Liu CP, Kao HL, Chao CL, et al. The relation of amino-terminal propeptide of type III procollagen and severity of coronary artery disease in patients without myocardial infarction or hibernation. Clin Biochem 2006;39:861–6.

[13] Henriksen NA, Mortensen JH, Sorensen LT, Bay-Jensen AC, Agren MS, Jorgensen LN, et al. The collagen turnover profile is altered in patients with inguinal and incisional hernia. Surgery 2015;157:312–21.

[14] El SA, Yano F, Mittal S, Filipi CJ. Collagen metabolism and recurrent hiatal hernia: cause and effect? Hernia 2006;10:511–20.

[15] Bourhis JM, Mariano N, Zhao Y, Harlos K, Exposito JY, Jones EY, et al. Structural basis of fibrillar collagen trimerization and related genetic disorders. Nat Struct Mol Biol 2012;19:1031–6.

[16] rteaga-Solis E, Gayraud B, Ramirez F. Elastic and collagenous networks in vascular diseases. Cell Struct Funct 2000;25:69–72.

[17] Smith LB, Hadoke PW, Dyer E, Denvir MA, Brownstein D, Miller E, et al. Haploinsufficiency of the murine Col3a1 locus causes aortic dissection: a novel model of the vascular type of Ehlers–Danlos syndrome. Cardiovasc Res 2011;90:182–90.

[18] Lisman T, Raynal N, Groeneveld D, Maddox B, Peachey AR, Huizinga EG, et al. A single high-affinity binding site for von Willebrand factor in collagen III, identified using synthetic triple-helical peptides. Blood 2006;108:3753–6.

[19] Luo R, Jeong SJ, Jin Z, Strokes N, Li S, Piao X. G protein-coupled receptor 56 and collagen III, a receptor-ligand pair, regulates cortical development and lamination. Proc Natl Acad Sci USA 2011;108:12925–30.

[20] Xu H, Raynal N, Stathopoulos S, Myllyharju J, Farndale RW, Leitinger B. Collagen binding specificity of the discoidin domain receptors: binding sites on collagens II and III and molecular determinants for collagen IV recognition by DDR1. Matrix Biol 2011;30:16–26.

[21] Vogel WF, Abdulhussein R, Ford CE. Sensing extracellular matrix: an update on discoidin domain receptor function. Cell Signal 2006;18:1108–16.

[22] Horslev-Petersen K. Circulating extracellular matrix components as markers for connective tissue response to inflammation. A clinical and experimental study with special emphasis on serum aminoterminal type III procollagen peptide in rheumatic diseases. Dan Med Bull 1990;37:308–29.

[23] Teppo AM, Tornroth T, Honkanen E, Gronhagen-Riska C. Urinary amino-terminal propeptide of type III procollagen (PIIINP) as a marker of interstitial fibrosis in renal transplant recipients. Transplantation 2003;75:2113–9.

[24] Parkes J, Guha IN, Roderick P, Harris S, Cross R, Manos MM, et al. Enhanced Liver Fibrosis (ELF) test accurately identifies liver fibrosis in patients with chronic hepatitis C. J Viral Hepat 2011;18:23–31.

[25] Oberti F, Valsesia E, Pilette C, Rousselet MC, Bedossa P, Aube C, et al. Noninvasive diagnosis of hepatic fibrosis or cirrhosis. Gastroenterology 1997;113:1609–16.

[26] Guechot J, Laudat A, Loria A, Serfaty L, Poupon R, Giboudeau J. Diagnostic accuracy of hyaluronan and type III procollagen amino-terminal peptide serum assays as markers of liver fibrosis in chronic viral hepatitis C evaluated by ROC curve analysis. Clin Chem 1996;42: 558–63.

[27] Fontana RJ, Dienstag JL, Bonkovsky HL, Sterling RK, Naishadham D, Goodman ZD, et al. Serum fibrosis markers are associated with liver disease progression in non-responder patients with chronic hepatitis C. Gut 2010;59:1401–9.

[28] Soylemezoglu O, Wild G, Dalley AJ, MacNeil S, Milford-Ward A, Brown CB, et al. Urinary and serum type III collagen: markers of renal fibrosis. Nephrol Dial Transplant 1997;12: 1883–9.

[29] Ghoul BE, Squalli T, Servais A, Elie C, Meas-Yedid V, Trivint C, et al. Urinary procollagen III aminoterminal propeptide (PIIINP): a fibrotest for the nephrologist. Clin J Am Soc Nephrol 2010;5:205–10.

[30] Barascuk N, Veidal SS, Larsen L, Larsen DV, Larsen MR, Wang J, et al. A novel assay for extracellular matrix remodeling associated with liver fibrosis: an enzyme-linked immunosorbent assay (ELISA) for a MMP-9 proteolytically revealed neo-epitope of type III collagen. Clin Biochem 2010;43:899–904.

[31] Nielsen MJ, Nedergaard AF, Sun S, Veidal SS, Larsen L, Zheng Q, et al. The neo-epitope specific PRO-C3 ELISA measures true formation of type III collagen associated with liver and muscle parameters. Am J Transl Res 2013;5:303–15.

[32] Karsdal MA, Delvin E, Christiansen C. Protein fingerprints–relying on and understanding the information of serological protein measurements. Clin Biochem 2011;44:1278–9.

[33] Wang WM, Ge G, Lim NH, Nagase H, Greenspan DS. TIMP-3 inhibits the procollagen N-proteinase ADAMTS-2. Biochem J 2006;398:515–9.

[34] Genovese F, Boor P, Papasotiriou M, Leeming DJ, Karsdal MA, Floege J. Turnover of type III collagen reflects disease severity and is associated with progression and microinflammation in patients with IgA nephropathy. Nephrol Dial Transplant 2016;31:472–79.

[35] Papasotiriou M, Genovese F, Klinkhammer BM, Kunter U, Nielsen SH, Karsdal MA, et al. Serum and urine markers of collagen degradation reflect renal fibrosis in experimental kidney diseases. Nephrol Dial Transplant 2015;30:1112–21.

[36] Leeming DJ, Karsdal MA, Byrjalsen I, Bendtsen F, Trebicka J, Nielsen MJ, et al. Novel serological neo-epitope markers of extracellular matrix proteins for the detection of portal hypertension. Aliment Pharmacol Ther 2013;38:1086–96.

[37] Leeming DJ, Anadol E, Schierwagen R, Karsdal MA, Byrjalsen I, Nielsen MJ, et al. Combined antiretroviral therapy attenuates hepatic extracellular matrix remodeling in HIV patients assessed by novel protein fingerprint markers. AIDS 2014;28:2081–90.

[38] Nielsen MJ, Kazankov K, Leeming DJ, Karsdal MA, Barrera F, McLeod D, et al. Markers of collagen remodeling detect clinically significant fibrosis in chronic hepatitis C patients. PLoS One 2015;10:e0137302.

[39] Nielsen MJ, Veidal SS, Karsdal MA, Orsnes-Leeming DJ, Vainer B, Gardner SD, et al. Plasma Pro-C3 (N-terminal type III collagen propeptide) predicts fibrosis progression in patients with chronic hepatitis C. Liver Int 2015;35:429–37.

[40] Schierwagen R, Leeming DJ, Klein S, Granzow M, Nielsen MJ, Sauerbruch T, et al. Serum markers of the extracellular matrix remodeling reflect antifibrotic therapy in bile-duct ligated rats. Front Physiol 2013;4:195.

[41] Veidal SS, Nielsen MJ, Leeming DJ, Karsdal MA. Phosphodiesterase inhibition mediates matrix metalloproteinase activity and the level of collagen degradation fragments in a liver fibrosis ex vivo rat model. BMC Res Notes 2012;5:686.

[42] Leeming DJ, Sand JM, Nielsen MJ, Genovese F, Martinez FJ, Hogaboam CM, et al. Serological investigation of the collagen degradation profile of patients with chronic obstructive pulmonary disease or idiopathic pulmonary fibrosis. Biomark Insights 2012;7:119–26.

[43] Jenkins RG, Simpson JK, Saini G, Bentley JH, Russell AM, Braybrooke R, et al. Longitudinal change in collagen degradation biomarkers in idiopathic pulmonary fibrosis: an analysis from the prospective, multicentre PROFILE study. Lancet Respir Med 2015;3:462–72.

[44] Sand JM, Knox AJ, Lange P, Sun S, Kristensen JH, Leeming DJ, et al. Accelerated extracellular matrix turnover during exacerbations of COPD. Respir Res 2015;16:69.

[45] rendt-Nielsen L, Eskehave TN, Egsgaard LL, Petersen KK, Graven-Nielsen T, Hoeck HC, et al. Association between experimental pain biomarkers and serologic markers in patients with different degrees of painful knee osteoarthritis. Arthritis Rheumatol 2014;66:3317–26.

[46] Bay-Jensen AC, Wichuk S, Byrjalsen I, Leeming DJ, Morency N, Christiansen C, et al. Circulating protein fragments of cartilage and connective tissue degradation are diagnostic and prognostic markers of rheumatoid arthritis and ankylosing spondylitis. PLoS One 2013;8:e54504.

[47] Bay-Jensen AC, Platt A, Byrjalsen I, Vergnoud P, Christiansen C, Karsdal MA. Effect of tocilizumab combined with methotrexate on circulating biomarkers of synovium, cartilage, and bone in the LITHE study. Semin Arthritis Rheum 2014;43:470–8.

[48] Kuivaniemi H, Tromp G, Prockop DJ. Mutations in fibrillar collagens (types I, II, III, and XI), fibril-associated collagen (type IX), and network-forming collagen (type X) cause a spectrum of diseases of bone, cartilage, and blood vessels. Hum Mutat 1997;9:300–15.

[49] Plancke A, Holder-Espinasse M, Rigau V, Manouvrier S, Claustres M, Khau Van KP. Homozygosity for a null allele of COL3A1 results in recessive Ehlers–Danlos syndrome. Eur J Hum Genet 2009;17:1411–6.

[50] Wang Q, Peng Z, Xiao S, Geng S, Yuan J, Li Z. RNAi-mediated inhibition of COL1A1 and COL3A1 in human skin fibroblasts. Exp Dermatol 2007;16:611–7.

[51] Gauglitz GG, Korting HC, Pavicic T, Ruzicka T, Jeschke MG. Hypertrophic scarring and keloids: pathomechanisms and current and emerging treatment strategies. Mol Med 2011;17:113–25.

[52] Bork P. The modular architecture of a new family of growth regulators related to connective tissue growth factor. FEBS Lett 1993;327:125–30.

[53] McLaughlin SH, Bulleid NJ. Molecular recognition in procollagen chain assembly. Matrix Biol 1998;16:369–77.

[54] Lebbink RJ, de Ruiter T, Adelmeijer J, Brenkman AB, van Helvoort JM, Koch M, et al. Collagens are functional, high affinity ligands for the inhibitory immune receptor LAIR-1. J Exp Med 2006;203:1419–25.

[55] Somasundaram R, Schuppan D. Type I, II, III, IV, V, and VI collagens serve as extracellular ligands for the isoforms of platelet-derived growth factor (AA, BB, and AB). J Biol Chem 1996;271:26884–91.

[56] Pope FM, Martin GR, Lichtenstein JR, Penttinen R, Gerson B, Rowe DW, et al. Patients with Ehlers–Danlos syndrome type IV lack type III collagen. Proc Natl Acad Sci USA 1975;72:1314–6.

[57] Zoppi N, Gardella R, De PA, Barlati S, Colombi M. Human fibroblasts with mutations in COL5A1 and COL3A1 genes do not organize collagens and fibronectin in the extracellular matrix, down-regulate alpha2beta1 integrin, and recruit alphavbeta3 Instead of alpha5beta1 integrin. J Biol Chem 2004;279:18157–68.
[58] Howard PS, Kucich U, Coplen DE, He Y. Transforming growth factor-beta1-induced hypertrophy and matrix expression in human bladder smooth muscle cells. Urology 2005;66: 1349–53.
[59] Brocks DG, Steinert C, Gerl M, Knolle J, Neubauer HP, Gunzler V. A radioimmunoassay for the N-terminal propeptide of rat procollagen type III. Application to the study of the uptake of the N-terminal propeptide of procollagen type III in isolated perfused rat liver. Matrix 1993;13:381–7.
[60] Vassiliadis E, Larsen DV, Clausen RE, Veidal SS, Barascuk N, Larsen L, et al. Measurement of CO3-610, a potential liver biomarker derived from matrix Metalloproteinase-9 degradation of collagen type III, in a rat model of reversible carbon-tetrachloride-induced fibrosis. Biomark Insights 2011;6:49–58.

Chapter 4

Type IV Collagen

J.M.B. Sand, F. Genovese, M.A. Karsdal
Nordic Bioscience, Herlev, Denmark

Chapter Outline

SUMMARY

Type IV collagen is the main collagen component of the basement membrane. It is a network-forming collagen that underlies epithelial and endothelial cells and functions as a barrier between tissue compartments. Type IV collagen has many binding partners and forms the backbone of the basement membrane. It holds important signaling potential as subdomains such as tumstatin are released when the protein is degraded by special proteases. Consequently, type IV collagen is both the most important structural collagen of the basement membrane and it entails key signaling potential, which is important for various physiological and pathological functions. The most well-studied mutations in type IV collagen cause Alport syndrome, a chronic kidney disease. Several biomarkers of type IV collagen have been developed, both formation and degradation fragments as well as whole domains such as 7S, documenting the importance of type IV collagen turnover in most, if not all, connective tissue diseases.

Type IV collagen is a crucial structural component of the basement membrane [1], a specialized sheet-like extracellular matrix (ECM) that underlies epithelial and endothelial cells and functions as a barrier between tissue compartments. Type IV collagen is a network-forming collagen that provides a molecular scaffold and interacts with cells, growth factors, and other basement membrane components such as laminin, nidogen, and perlecan [2–7]. The specific interactions result in a specialized basement membrane, unique for each tissue, which is involved in several important biological processes including cell adhesion, migration, development, tissue regeneration, and wound healing [8]. Six different α chains of type IV collagen, α1–α6, are expressed in vertebrates [9]. They are encoded by the genes COL4A1 to COL4A6, which are organized into pairs in a head-to-head manner [9–12].

Each α chain consists of a short N-terminal collagenous 7S domain, a central collagenous domain of approximately 1400 residues, and a C-terminal globular non-collagenous (NC) 1 domain of approximately 230 residues [8]. The collagenous domain of type IV collagen differs from that found in the fibrillar collagens as it contains 21–26 characteristic interruptions of the proline-rich Gly-Xaa-Yaa triple repeat [8]. These frequent interruptions result in a less rigid structure and provide type IV collagen with extensive flexibility that enables the formation of a network and may allow for specific proteolytic cleavages of the triple helix or serve as sites of cell binding and interchain crosslinking [13]. Three α chains self-assemble into a heterotrimeric protomer by the association of their NC1 domains and the subsequent supercoiling of the triple helical domains in a zipper-like manner. The association of the NC1 domains is stabilized by domain-swapping interactions [14], and the presence of recognition sequences ensures the formation of only three different protomer isoforms: α1α1α2(IV), α3α4α5(IV), and α5α5α6(IV) [15]. After secretion to the ECM, the protomers self-assemble into complex lattice-shaped networks. The NC1 domain of two protomers associate end-to-end, forming an NC1 hexamer, and the 7S domains of four protomers form a 7S tetramer [16,17]. The 7S domain is rich in cysteine and lysine residues that stabilize the 7S tetramer by forming disulfide bonds and lysine–hydroxylysine crosslinks. Furthermore, the 7S tetramer is heavily glycosylated, making it resistant to collagenase activity [18]. The NC1 hexamer, although containing conserved cysteine residues, has been shown to be stabilized by covalent Met–Lys crosslinks with only intrachain disulfide bonds [14,19,20].

The type IV collagen networks consist of α1α1α2(IV), α3α4α5(IV), or a combination of α1α1α2(IV) and α5α5α6(IV) connected through their NC1 domains [21]. The predominant network α1α1α2(IV) is found in basement membranes of all tissues, whereas the other networks show restricted tissue distribution. α3α4α5(IV) has been identified in the glomerular basement membrane of the kidney and in the alveolar basement membrane of the lung as well as in basement membranes of testis, inner ear cochlea, and eye, whereas α5α5α6(IV) is expressed in the basement membrane of the bronchial epithelium, smooth muscle cells, skin, and Bowman capsule and tubular basement membrane of the kidney [8,15,22–26]. Throughout embryonic development, the ubiquitous α1α1α2(IV) is replaced by α3α4α5(IV) and α5α5α6(IV) in specific tissues in a developmental switch [24]. In contrast to the α1α1α2(IV) network, the α3α4α5(IV) has a high degree of crosslinking and resistance to proteolysis [27] that may enable it to withstand increased stress in locations such as the mature glomerular and alveolar basement membranes.

Inherited mutations in the genes encoding different type IV collagen α chains have been implicated in several severe diseases mainly showing cerebrovascular and renal manifestations. Mutations in the genes encoding the α1(IV) and α2(IV) chains are mainly expressed as vascular defects resulting in, for example, intracerebral hemorrhages and intracranial aneurysms [28].

Patients with hereditary angiopathy with nephropathy, aneurysm, and muscle cramps syndrome show mutations in the COL4A1 gene, causing a systemic phenotype due to the widespread distribution of the α1(IV) chain and involving both renal, vascular, and ocular manifestations [29]. Alport syndrome is caused by mutations in the α3(IV), α4(IV), or α5(IV) chain, resulting in the absence of the α3α4α5(IV) network. Several variations of Alport syndrome exist, but classically it consists of hematuria, proteinuria, progressive renal failure, and sensorineural deafness [30]. Patients retain a fetal distribution of the α1α1α2(IV) network instead of switching to α3α4α5(IV) during development, leaving especially the glomerular basement membrane susceptible to proteolytic degradation [27]. Type IV collagen is the target of two autoimmune diseases affecting the kidney: Goodpasture syndrome and Alport posttransplantation disease. Both diseases are characterized by autoantibodies attacking the glomerular basement membrane and causing rapidly progressive glomerulonephritis [31]. In addition, patients with Goodpasture syndrome present with lung hemorrhage as the autoantibodies also attack the alveolar basement membrane. These insults result from autoantibodies directed at cryptic epitopes in the NC1 domain of the α3(IV) chain, which are present in both the glomerular and alveolar basement membranes [30]. The cryptic sites are inaccessible in the native protein conformation and only become accessible to autoantibodies after the dissociation of the NC1 hexamer caused by, for example, oxidative stress [32].

A specific 28-kDa peptide from the NC1 domain of the α3(IV) chain known as tumstatin have anti-angiogenic properties and is able to suppress tumor growth [33]. Matrix metalloproteinase (MMP)-9 controls the physiological levels of tumstatin by cleaving it from type IV collagen in the basement membrane [33]. The inhibitory effects of tumstatin on tumor growth have been studied for several years. Tumstatin can directly bind to the tumor endothelium, and its anti-angiogenic action is carried out through the interaction with αvβ3 integrin on proliferating endothelial cells [34]. Furthermore, tumstatin levels in the airways of patients with asthma are reduced, which may contribute to disease progression as tumstatin suppresses airway angiogenesis, hyperresponsiveness, and inflammation [35]. The use of tumstatin as a therapeutic agent has been suggested for diabetic nephropathy [36] and cancer [37]. Tumstatin can also be a diagnostic, prognostic, and therapeutic marker of renal carcinoma [38]. Further investigations of fragments of type IV collagen have revealed that not only tumstatin but also several other peptides released from type IV collagen may exert biological activities that are different from those of the intact protein. Such peptides deriving from the ECM are known as matrikines. Matrikines derived from the NC1 domain of both the α1 (arresten), α2 (canstatin), α3 (tumstatin), α4 (tetrastatin), α5 (pentastatin), and α6 (hexastatin) chains of type IV collagen have been identified and shown to inhibit tumor progression by, for example, anti-angiogenic effects [33,39–41]. Interestingly, the work on matrikines indicates that fragments of type IV collagen can be more than just products of disease, but may play an active role in suppressing or sustaining a pathological state.

BIOMARKERS OF TYPE IV COLLAGEN

The tissue-specific nature of the α3-α6(IV) chains introduces the opportunity to use selected peptides of type IV collagen as tissue-specific biomarkers of disease. Domains and fragments originating from specific α chains of type IV collagen released as a result of protein remodeling have been widely investigated for their use as biomarkers of various diseases of the ECM. In addition to the matrikines, specific fragments of type IV collagen produced by MMPs in relation to protein degradation have been investigated for their potential as biomarkers of fibroproliferative diseases. Circulating levels of an MMP-generated fragment of the α1 chain, C4M, have been associated with the presence of ankylosing spondylitis [42], liver fibrosis [43–45], portal hypertension [46], pancreatic cancer [47], and chronic obstructive pulmonary disease (COPD) [48,49]. Interestingly, C4M has also been found to be elevated in serum of patients with COPD having an acute exacerbation compared to stable COPD, indicating an association with disease activity [49]. Elevated circulating levels of the 7S domain, believed to reflect type IV collagen formation, have been associated with fibrosis of the liver [50–52] and lung [53] as well as an increased mortality risk in patients undergoing hemodialysis [54].

Type IV collagen is also a biomarker target in urine of patients with different kidney disease manifestations. In type 1 diabetic nephropathy, the estimated glomerular filtration rate declined more rapidly in patients with an elevated urinary type IV collagen-to-creatinine ratio (T4C), but with normal albumin-to-creatinine ratio, than patients with normal T4C [55]. A proteomic study observing the levels of ECM protein fragments in patients with type 1 diabetes presenting a progressive early decline in renal function showed a decreased expression of urinary fragments of the α1(IV) chain compared with control subjects with stable renal function [56]. In type 2 diabetic patients, increased type IV collagen urine excretion was associated with the severity of morphological alterations in kidney fibrosis [57]. Moreover, elevated levels of urinary type IV collagen were correlated with urinary proteins, urinary *N*-acetyl-β-D-glucosaminidase, and selectivity index in a study on biopsy-proven membranous nephropathy and anti-neutrophil cytoplasmic antibodies–associated glomerulonephritis [58].

Type IV Collagen	Description	References
Gene name and number	COL4A1 (13q34); COL4A2 (13q34); COL4A3 (2q36-q37); COL4A4 (2q35-q36); COL4A5 (Xq22); COL4A6 (Xq22);	NCBI gene ID: 1282, 1284, 1285, 1286, 1287, and 1288

Type IV Collagen	Description	References
Mutations with diseases in humans	• COL4A1: familial porencephaly, intracerebral hemorrhage, hemorrhagic stroke, small-vessel disease, intracranial aneurysms, HANAC syndrome, retinal artery tortuosity. • COL4A2: familial porencephaly, intracerebral hemorrhage, hemorrhagic stroke, small-vessel disease. • COL4A3: autosomal recessive Alport syndrome, familial benign hematuria. • COL4A3/COL4A4: focal segmental glomerulosclerosis and renal failure in thin basement membrane nephropathy. • COL4A4: autosomal dominant Alport syndrome, familial benign hematuria. • COL4A5: X-linked Alport syndrome, diffuse leiomyomatosis. • COL4A6: diffuse leiomyomatosis, X-linked non-syndromic hearing loss.	[28,29,59–69]
Null mutation in mice	• COL4A1 KO, COL4A2 KO, and COL4A1/COL4A2 double KO are embryonically lethal. • COL4A3 KO is a model for autosomal Alport syndrome.	[70–72]
Tissue distribution in healthy states	• $\alpha1\alpha1\alpha2$(IV): basement membranes of all tissues. • $\alpha3\alpha4\alpha5$(IV): basement membranes of the glomeruli of the kidney; alveoli of the lungs, testis, inner ear cochlea, and eyes. • $\alpha5\alpha5\alpha6$(IV): basement membranes of bronchial epithelium, smooth muscle cells, skin, and Bowman capsule and tubules of the kidney.	[8,15,22–26]
Tissue distribution in pathological affected states	• Alport syndrome: loss of the $\alpha3\alpha4\alpha5$(IV) network in kidneys and lungs. • Tumors: loss of $\alpha3$–$\alpha6$ chains; loss or accumulation of $\alpha1$ and $\alpha2$, depending on tumor type. • Fibrosis: accumulation in liver and kidney, disruption in lung.	[22,30,73–78]
Special domains	N-terminal 7S domain; collagenous domain with 21–26 interruptions; C-terminal NC1 domain	[8]
Special neoepitopes	• NC1 domain of $\alpha1$ (arresten), $\alpha2$ (canstatin), $\alpha3$ (tumstatin), $\alpha4$ (tetrastatin), $\alpha5$ (pentastatin), and $\alpha6$ (hexastatin). • 7S domain. • Degradation fragments (eg, C4M)	[18,33,39–41, 43,44]
Protein structure and function	Three α chains form the heterotrimeric protomers $\alpha1\alpha1\alpha2$(IV), $\alpha3\alpha4\alpha5$(IV), and $\alpha5\alpha5\alpha6$(IV). Protomers associate via two NC1 and four 7S domains to form lattice-shaped networks in basement membranes.	[17]
Binding proteins	Integrins, DDR1, nidogen, laminin, heparan sulfate proteoglycan, fibronectin, TGF-β, PDGF.	[2–7]

Type IV Collagen	Description	References
Known central function	Underlies endothelial and epithelial cells, separates tissue compartments; surrounds various cells including smooth muscle and nerve cells; functions in cell adhesion, migration and development, tissue regeneration, and wound healing.	[8]
Animals models with protein affected	• COL4A3 KO mouse model for autosomal Alport syndrome • COL4A4 splice site mutation resulting in mutant α4 chain expression in tissue as mouse model of autosomal Alport syndrome • COL4A5 nonsense mutation resulting in male KO and female mosaic for α5(IV) chain expression as mouse model of X-linked Alport syndrome	[72,79,80]
Biomarkers	C4M, C4Ma3, P4NP 7S, 7S domain, NC1 domain (including arresten, canstatin, tumstatin, tetrastatin, pentastatin, hexastatin).	[33,39–41, 43,44,51,52, 81,82]

COL, Collagen; *DDR*, discoidin domain receptor; *HANAC*, hereditary angiopathy with nephropathy, aneurysms, and muscle cramps; *KO*, knockout; *NC*, non-collagenous; *NCBI*, National Center for Biotechnology Information; *PDGF*, platelet-derived growth factor; *TGF*, transforming growth factor.

REFERENCES

[1] Laurie GW, Leblond CP, Martin GR. Localization of type IV collagen, laminin, heparan sulfate proteoglycan, and fibronectin to the basal lamina of basement membranes. J Cell Biol 1982;95:340–4.

[2] Eble JA, Golbik R, Mann K, Kuhn K. The alpha 1 beta 1 integrin recognition site of the basement membrane collagen molecule [alpha 1(IV)]2 alpha 2(IV). EMBO J 1993;12: 4795–802.

[3] Xu H, Raynal N, Stathopoulos S, Myllyharju J, Farndale RW, Leitinger B. Collagen binding specificity of the discoidin domain receptors: binding sites on collagens II and III and molecular determinants for collagen IV recognition by DDR1. Matrix Biol 2011;30:16–26.

[4] Fox JW, Mayer U, Nischt R, Aumailley M, Reinhardt D, Wiedemann H, et al. Recombinant nidogen consists of three globular domains and mediates binding of laminin to collagen type IV. EMBO J 1991;10:3137–46.

[5] Paralkar VM, Vukicevic S, Reddi AH. Transforming growth factor beta type 1 binds to collagen IV of basement membrane matrix: implications for development. Dev Biol 1991;143: 303–8.

[6] Somasundaram R, Schuppan D. Type I, II, III, IV, V, and VI collagens serve as extracellular ligands for the isoforms of platelet-derived growth factor (AA, BB, and AB). J Biol Chem 1996;271:26884–91.

[7] Laurie GW, Bing JT, Kleinman HK, Hassell JR, Aumailley M, Martin GR, et al. Localization of binding sites for laminin, heparan sulfate proteoglycan and fibronectin on basement membrane (type IV) collagen. J Mol Biol 1986;189:205–16.

[8] Khoshnoodi J, Pedchenko V, Hudson BG. Mammalian collagen IV. Microsc Res Tech 2008;71:357–70.

[9] Zhou J, Ding M, Zhao Z, Reeders ST. Complete primary structure of the sixth chain of human basement membrane collagen, alpha 6(IV). Isolation of the cDNAs for alpha 6(IV) and comparison with five other type IV collagen chains. J Biol Chem 1994;269:13193–9.
[10] Soininen R, Huotari M, Hostikka SL, Prockop DJ, Tryggvason K. The structural genes for alpha 1 and alpha 2 chains of human type IV collagen are divergently encoded on opposite DNA strands and have an overlapping promoter region. J Biol Chem 1988;263:17217–20.
[11] Poschl E, Pollner R, Kuhn K. The genes for the alpha 1(IV) and alpha 2(IV) chains of human basement membrane collagen type IV are arranged head-to-head and separated by a bidirectional promoter of unique structure. EMBO J 1988;7:2687–95.
[12] Mariyama M, Zheng K, Yang-Feng TL, Reeders ST. Colocalization of the genes for the alpha 3(IV) and alpha 4(IV) chains of type IV collagen to chromosome 2 bands q35-q37. Genomics 1992;13:809–13.
[13] Vandenberg P, Kern A, Ries A, Luckenbill-Edds L, Mann K, Kuhn K. Characterization of a type IV collagen major cell binding site with affinity to the alpha 1 beta 1 and the alpha 2 beta 1 integrins. J Cell Biol 1991;113:1475–83.
[14] Sundaramoorthy M, Meiyappan M, Todd P, Hudson BG. Crystal structure of NC1 domains. Structural basis for type IV collagen assembly in basement membranes. J Biol Chem 2002;277:31142–53.
[15] Boutaud A, Borza DB, Bondar O, Gunwar S, Netzer KO, Singh N, et al. Type IV collagen of the glomerular basement membrane. Evidence that the chain specificity of network assembly is encoded by the noncollagenous NC1 domains. J Biol Chem 2000;275:30716–24.
[16] Oberbaumer I, Wiedemann H, Timpl R, Kuhn K. Shape and assembly of type IV procollagen obtained from cell culture. EMBO J 1982;1:805–10.
[17] Timpl R, Wiedemann H, van Delden V, Furthmayr H, Kuhn K. A network model for the organization of type IV collagen molecules in basement membranes. Eur J Biochem 1981;120:203–11.
[18] Risteli J, Bachinger HP, Engel J, Furthmayr H, Timpl R. 7-S collagen: characterization of an unusual basement membrane structure. Eur J Biochem 1980;108:239–50.
[19] Than ME, Henrich S, Huber R, Ries A, Mann K, Kuhn K, et al. The 1.9-A crystal structure of the noncollagenous (NC1) domain of human placenta collagen IV shows stabilization via a novel type of covalent Met-Lys cross-link. Proc Natl Acad Sci USA 2002;99:6607–12.
[20] Than ME, Bourenkov GP, Henrich S, Mann K, Bode W. The NC1 dimer of human placental basement membrane collagen IV: does a covalent crosslink exist? Biol Chem 2005;386:759–66.
[21] Borza DB, Bondar O, Ninomiya Y, Sado Y, Naito I, Todd P, et al. The NC1 domain of collagen IV encodes a novel network composed of the alpha 1, alpha 2, alpha 5, and alpha 6 chains in smooth muscle basement membranes. J Biol Chem 2001;276:28532–40.
[22] Nakano KY, Iyama KI, Mori T, Yoshioka M, Hiraoka T, Sado Y, et al. Loss of alveolar basement membrane type IV collagen alpha3, alpha4, and alpha5 chains in bronchioloalveolar carcinoma of the lung. J Pathol 2001;194:420–7.
[23] Ninomiya Y, Kagawa M, Iyama K, Naito I, Kishiro Y, Seyer JM, et al. Differential expression of two basement membrane collagen genes, COL4A6 and COL4A5, demonstrated by immunofluorescence staining using peptide-specific monoclonal antibodies. J Cell Biol 1995;130:1219–29.
[24] Miner JH, Sanes JR. Collagen IV alpha 3, alpha 4, and alpha 5 chains in rodent basal laminae: sequence, distribution, association with laminins, and developmental switches. J Cell Biol 1994;127:879–91.
[25] Kalluri R, Gattone VH, Hudson BG. Identification and localization of type IV collagen chains in the inner ear cochlea. Connect Tissue Res 1998;37:143–50.

[26] Gunwar S, Ballester F, Noelken ME, Sado Y, Ninomiya Y, Hudson BG. Glomerular basement membrane. Identification of a novel disulfide-cross-linked network of alpha3, alpha4, and alpha5 chains of type IV collagen and its implications for the pathogenesis of Alport syndrome. J Biol Chem 1998;273:8767–75.

[27] Kalluri R, Shield CF, Todd P, Hudson BG, Neilson EG. Isoform switching of type IV collagen is developmentally arrested in X-linked Alport syndrome leading to increased susceptibility of renal basement membranes to endoproteolysis. J Clin Invest 1997;99:2470–8.

[28] Gould DB, Phalan FC, van Mil SE, Sundberg JP, Vahedi K, Massin P, et al. Role of COL4A1 in small-vessel disease and hemorrhagic stroke. N Engl J Med 2006;354:1489–96.

[29] Plaisier E, Gribouval O, Alamowitch S, Mougenot B, Prost C, Verpont MC, et al. COL4A1 mutations and hereditary angiopathy, nephropathy, aneurysms, and muscle cramps. N Engl J Med 2007;357:2687–95.

[30] Hudson BG, Tryggvason K, Sundaramoorthy M, Neilson EG. Alport's syndrome, Goodpasture's syndrome, and type IV collagen. N Engl J Med 2003;348:2543–56.

[31] Chen YM, Miner JH. Glomerular basement membrane and related glomerular disease. Transl Res 2012;160:291–7.

[32] Borza DB, Netzer KO, Leinonen A, Todd P, Cervera J, Saus J, et al. The Goodpasture autoantigen. Identification of multiple cryptic epitopes on the NC1 domain of the alpha3(IV) collagen chain. J Biol Chem 2000;275:6030–7.

[33] Hamano Y, Zeisberg M, Sugimoto H, Lively JC, Maeshima Y, Yang C, et al. Physiological levels of tumstatin, a fragment of collagen IV alpha3 chain, are generated by MMP-9 proteolysis and suppress angiogenesis via alphaV beta3 integrin. Cancer Cell 2003;3:589–601.

[34] Eikesdal HP, Sugimoto H, Birrane G, Maeshima Y, Cooke VG, Kieran M, et al. Identification of amino acids essential for the antiangiogenic activity of tumstatin and its use in combination antitumor activity. Proc Natl Acad Sci USA 2008;105:15040–5.

[35] Burgess JK, Boustany S, Moir LM, Weckmann M, Lau JY, Grafton K, et al. Reduction of tumstatin in asthmatic airways contributes to angiogenesis, inflammation, and hyperresponsiveness. Am J Respir Crit Care Med 2010;181:106–15.

[36] Yamamoto Y, Maeshima Y, Kitayama H, Kitamura S, Takazawa Y, Sugiyama H, et al. Tumstatin peptide, an inhibitor of angiogenesis, prevents glomerular hypertrophy in the early stage of diabetic nephropathy. Diabetes 2004;53:1831–40.

[37] Li YJ, Sun LC, He Y, Liu XH, Liu M, Wang QM, et al. The anti-tumor properties of two tumstatin peptide fragments in human gastric carcinoma. Acta Pharmacol Sin 2009;30:1307–15.

[38] Xu CX, Liu XX, Hou GS, Yan YF, Chen SM, Wang W, et al. The expression of tumstatin is down-regulated in renal carcinoma. Mol Biol Rep 2010;37:2273–7.

[39] Colorado PC, Torre A, Kamphaus G, Maeshima Y, Hopfer H, Takahashi K, et al. Anti-angiogenic cues from vascular basement membrane collagen. Cancer Res 2000;60:2520–6.

[40] Kamphaus GD, Colorado PC, Panka DJ, Hopfer H, Ramchandran R, Torre A, et al. Canstatin, a novel matrix-derived inhibitor of angiogenesis and tumor growth. J Biol Chem 2000;275:1209–15.

[41] Karagiannis ED, Popel AS. Identification of novel short peptides derived from the alpha 4, alpha 5, and alpha 6 fibrils of type IV collagen with anti-angiogenic properties. Biochem Biophys Res Commun 2007;354:434–9.

[42] Bay-Jensen AC, Leeming DJ, Kleyer A, Veidal SS, Schett G, Karsdal MA. Ankylosing spondylitis is characterized by an increased turnover of several different metalloproteinase-derived collagen species: a cross-sectional study. Rheumatol Int 2012;32:3565–72.

[43] Veidal SS, Karsdal MA, Nawrocki A, Larsen MR, Dai Y, Zheng Q, et al. Assessment of proteolytic degradation of the basement membrane: a fragment of type IV collagen as a biochemical marker for liver fibrosis. Fibrogenesis Tissue Repair 2011;4:22.

[44] Sand JM, Larsen L, Hogaboam C, Martinez F, Han M, Rossel LM, et al. MMP mediated degradation of type IV collagen alpha 1 and alpha 3 chains reflects basement membrane remodeling in experimental and clinical fibrosis–validation of two novel biomarker assays. PLoS One 2013;8:e84934.

[45] Leeming DJ, Byrjalsen I, Jimenez W, Christiansen C, Karsdal MA. Protein fingerprinting of the extracellular matrix remodelling in a rat model of liver fibrosis–a serological evaluation. Liver Int 2013;33:439–47.

[46] Leeming DJ, Karsdal MA, Byrjalsen I, Bendtsen F, Trebicka J, Nielsen MJ, et al. Novel serological neo-epitope markers of extracellular matrix proteins for the detection of portal hypertension. Aliment Pharmacol Ther 2013;38:1086–96.

[47] Willumsen N, Bager CL, Leeming DJ, Smith V, Karsdal MA, Dornan D, et al. Extracellular matrix specific protein fingerprints measured in serum can separate pancreatic cancer patients from healthy controls. BMC Cancer 2013;13:554.

[48] Leeming DJ, Sand JM, Nielsen MJ, Genovese F, Martinez FJ, Hogaboam CM, et al. Serological investigation of the collagen degradation profile of patients with chronic obstructive pulmonary disease or idiopathic pulmonary fibrosis. Biomark Insights 2012;7:119–26.

[49] Sand JM, Knox AJ, Lange P, Sun S, Kristensen JH, Leeming DJ, et al. Accelerated extracellular matrix turnover during exacerbations of COPD. Respir Res 2015;16:69.

[50] Murawaki Y, Ikuta Y, Koda M, Yamada S, Kawasaki H. Comparison of serum 7S fragment of type IV collagen and serum central triple-helix of type IV collagen for assessment of liver fibrosis in patients with chronic viral liver disease. J Hepatol 1996;24:148–54.

[51] Bentsen KD, Horn T, Risteli J, Risteli L, Engstrom-Laurent A, Horslev-Petersen K, et al. Serum aminoterminal type III procollagen peptide and the 7S domain of type IV collagen in patients with alcohol abuse. Relation to ultrastructural fibrosis in the acinar zone 3 and to serum hyaluronan. Liver 1987;7:339–46.

[52] Leeming DJ, Nielsen MJ, Dai Y, Veidal SS, Vassiliadis E, Zhang C, et al. Enzyme-linked immunosorbent serum assay specific for the 7S domain of Collagen Type IV (P4NP 7S): a marker related to the extracellular matrix remodeling during liver fibrogenesis. Hepatol Res 2012;42(5):482–93.

[53] Kasuga I, Yonemaru M, Kiyokawa H, Ichinose Y, Toyama K. Clinical evaluation of serum type IV collagen 7S in idiopathic pulmonary fibrosis. Respirology 1996;1:277–81.

[54] Leeming DJ, Karsdal MA, Rasmussen LM, Scholze A, Tepel M. Association of systemic collagen type IV formation with survival among patients undergoing hemodialysis. PLoS One 2013;8:e71050.

[55] Morita M, Uchigata Y, Hanai K, Ogawa Y, Iwamoto Y. Association of urinary type IV collagen with GFR decline in young patients with type 1 diabetes. Am J Kidney Dis 2011;58:915–20.

[56] Merchant ML, Perkins BA, Boratyn GM, Ficociello LH, Wilkey DW, Barati MT, et al. Urinary peptidome may predict renal function decline in type 1 diabetes and microalbuminuria. J Am Soc Nephrol 2009;20:2065–74.

[57] Okonogi H, Nishimura M, Utsunomiya Y, Hamaguchi K, Tsuchida H, Miura Y, et al. Urinary type IV collagen excretion reflects renal morphological alterations and type IV collagen expression in patients with type 2 diabetes mellitus. Clin Nephrol 2001;55:357–64.

[58] Furumatsu Y, Nagasawa Y, Shoji T, Yamamoto R, Iio K, Matsui I, et al. Urinary type IV collagen in nondiabetic kidney disease. Nephron Clin Pract 2011;117:c160–6.

[59] Breedveld G, de Coo IF, Lequin MH, Arts WF, Heutink P, Gould DB, et al. Novel mutations in three families confirm a major role of COL4A1 in hereditary porencephaly. J Med Genet 2006;43:490–5.

[60] Jeanne M, Labelle-Dumais C, Jorgensen J, Kauffman WB, Mancini GM, Favor J, et al. COL4A2 mutations impair COL4A1 and COL4A2 secretion and cause hemorrhagic stroke. Am J Hum Genet 2012;90:91–101.
[61] Verbeek E, Meuwissen ME, Verheijen FW, Govaert PP, Licht DJ, Kuo DS, et al. COL4A2 mutation associated with familial porencephaly and small-vessel disease. Eur J Hum Genet 2012;20:844–51.
[62] Lemmink HH, Mochizuki T, van den Heuvel LP, Schroder CH, Barrientos A, Monnens LA, et al. Mutations in the type IV collagen alpha 3 (COL4A3) gene in autosomal recessive Alport syndrome. Hum Mol Genet 1994;3:1269–73.
[63] Mochizuki T, Lemmink HH, Mariyama M, Antignac C, Gubler MC, Pirson Y, et al. Identification of mutations in the alpha 3(IV) and alpha 4(IV) collagen genes in autosomal recessive Alport syndrome. Nat Genet 1994;8:77–81.
[64] Rana K, Tonna S, Wang YY, Sin L, Lin T, Shaw E, et al. Nine novel COL4A3 and COL4A4 mutations and polymorphisms identified in inherited membrane diseases. Pediatr Nephrol 2007;22:652–7.
[65] Lemmink HH, Nillesen WN, Mochizuki T, Schroder CH, Brunner HG, van Oost BA, et al. Benign familial hematuria due to mutation of the type IV collagen alpha4 gene. J Clin Invest 1996;98:1114–8.
[66] Barker DF, Hostikka SL, Zhou J, Chow LT, Oliphant AR, Gerken SC, et al. Identification of mutations in the COL4A5 collagen gene in Alport syndrome. Science 1990;248:1224–7.
[67] Zhou J, Mochizuki T, Smeets H, Antignac C, Laurila P, de Paepe A, et al. Deletion of the paired alpha 5(IV) and alpha 6(IV) collagen genes in inherited smooth muscle tumors. Science 1993;261:1167–9.
[68] Rost S, Bach E, Neuner C, Nanda I, Dysek S, Bittner RE, et al. Novel form of X-linked nonsyndromic hearing loss with cochlear malformation caused by a mutation in the type IV collagen gene COL4A6. Eur J Hum Genet 2014;22:208–15.
[69] Voskarides K, Damianou L, Neocleous V, Zouvani I, Christodoulidou S, Hadjiconstantinou V, et al. COL4A3/COL4A4 mutations producing focal segmental glomerulosclerosis and renal failure in thin basement membrane nephropathy. J Am Soc Nephrol 2007;18: 3004–16.
[70] Favor J, Gloeckner CJ, Janik D, Klempt M, Neuhauser-Klaus A, Pretsch W, et al. Type IV procollagen missense mutations associated with defects of the eye, vascular stability, the brain, kidney function and embryonic or postnatal viability in the mouse, *Mus musculus*: an extension of the Col4a1 allelic series and the identification of the first two Col4a2 mutant alleles. Genetics 2007;175:725–36.
[71] Poschl E, Schlotzer-Schrehardt U, Brachvogel B, Saito K, Ninomiya Y, Mayer U. Collagen IV is essential for basement membrane stability but dispensable for initiation of its assembly during early development. Development 2004;131:1619–28.
[72] Cosgrove D, Meehan DT, Grunkemeyer JA, Kornak JM, Sayers R, Hunter WJ, et al. Collagen COL4A3 knockout: a mouse model for autosomal Alport syndrome. Genes Dev 1996;10:2981–92.
[73] Gubler MC, Knebelmann B, Beziau A, Broyer M, Pirson Y, Haddoum F, et al. Autosomal recessive Alport syndrome: immunohistochemical study of type IV collagen chain distribution. Kidney Int 1995;47:1142–7.
[74] Hiki Y, Iyama K, Tsuruta J, Egami H, Kamio T, Suko S, et al. Differential distribution of basement membrane type IV collagen alpha1(IV), alpha2(IV), alpha5(IV) and alpha6(IV) chains in colorectal epithelial tumors. Pathol Int 2002;52:224–33.

[75] Oka Y, Naito I, Manabe K, Sado Y, Matsushima H, Ninomiya Y, et al. Distribution of collagen type IV alpha1-6 chains in human normal colorectum and colorectal cancer demonstrated by immunofluorescence staining using chain-specific epitope-defined monoclonal antibodies. J Gastroenterol Hepatol 2002;17:980–6.
[76] Hahn E, Wick G, Pencev D, Timpl R. Distribution of basement membrane proteins in normal and fibrotic human liver: collagen type IV, laminin, and fibronectin. Gut 1980;21:63–71.
[77] Sharma AK, Mauer SM, Kim Y, Michael AF. Interstitial fibrosis in obstructive nephropathy. Kidney Int 1993;44:774–88.
[78] Raghu G, Striker LJ, Hudson LD, Striker GE. Extracellular matrix in normal and fibrotic human lungs. Am Rev Respir Dis 1985;131:281–9.
[79] Korstanje R, Caputo CR, Doty RA, Cook SA, Bronson RT, Davisson MT, et al. A mouse Col4a4 mutation causing Alport glomerulosclerosis with abnormal collagen alpha3alpha4alpha5(IV) trimers. Kidney Int 2014;85:1461–8.
[80] Rheault MN, Kren SM, Thielen BK, Mesa HA, Crosson JT, Thomas W, et al. Mouse model of X-linked Alport syndrome. J Am Soc Nephrol 2004;15:1466–74.
[81] Risteli J, Rohde H, Timpl R. Sensitive radioimmunoassays for 7 S collagen and laminin: application to serum and tissue studies of basement membranes. Anal Biochem 1981;113:372–8.
[82] Schuppan D, Besser M, Schwarting R, Hahn EG. Radioimmunoassay for the carboxy-terminal cross-linking domain of type IV (basement membrane) procollagen in body fluids. Characterization and application to collagen type IV metabolism in fibrotic liver disease. J Clin Invest 1986;78:241–8.

Chapter 5

Type V Collagen

D.J. Leeming, M.A. Karsdal
Nordic Bioscience, Herlev, Denmark

Chapter Outline

SUMMARY

Type V collagen is a fibrillar collagen. Type V collagen is essential for fibrillation of types I and III collagen, and consequently for optimal fibrillary formation and tissue quality. Type V collagen contributes to the bone matrix; corneal stroma; and the interstitial matrix of muscles, liver, lungs, and placenta. Dysregulation of collagen fibrillogenesis is a hallmark of several subtypes of Ehlers–Danlos syndrome (EDS). Around 50% of patients with classic EDS harbor a heterozygous mutation in type V collagen, causing connective diseases as a result of decreased matrix quality. Collagen type V knockout mice synthesize and secrete normal amounts of type I collagen, but collagen fibrils are virtually absent and mice die at the onset of organogenesis, indicating that normal fibrillogenesis, such as that produced by type V collagen, is crucial. Several biomarkers of type V collagen formation and degradation exist, which have proven type V collagen to be of special importance in connective tissue diseases.

Type V collagen is also a member of group I collagens known as fibrillar-forming collagens and is ubiquitously distributed throughout the body. Type V collagen binds to DNA, heparan sulfate, thrombospondin, heparin, and insulin. It is a key determinant in the assembly of tissue-specific collagen matrices such as those formed by types I and III. Type V collagen exists as α1, α2, and α3 chains, and each structure contains two propeptides, P5NP (at the N terminus) and P5CP (at the C terminus). The most abundant isoform of type V collagen is the [α1(V)]2α2(V) heterotrimer [1]. During fibril formation, this heterotrimer is incorporated into type I collagen fibrils and in doing so increases the diameter of the fibrils by partial retention of the α1(V)-N-propeptide [1,2]. It has been proposed that the total type V collagen triple helix is buried within the fibril, in contrast to type I collagen, which is present along the fibril surface. The α1(V)-N-propeptide extends outward from the gap zones of the heterotypic fibrils and is thus the only part of the type V collagen molecule that emerges at the surface of these fibrils [1].

A characteristic of type V collagen is the partial retention of the N-propeptide extension. In its normal healthy state type V collagen is resistant to mammalian collagenases and sensitive to trypsin [3]. It contributes to the organic bone matrix; corneal stroma; and the interstitial matrix of muscles, liver, lungs, and placenta [3]. Collagen knockout mice (col5α1 –/–) synthesize and secrete normal amounts of type I collagen, but collagen fibrils are virtually absent and mice die at the onset of organogenesis [4], indicating that normal fibrillogenesis, such as that produced by type V collagen, is crucial.

Ehlers–Danlos syndrome (EDS) is a group of heritable connective tissue disorders that share common features of skin hyperextensibility, articular hypermobility, and tissue fragility [5]. Dysregulation of collagen fibrillogenesis is a hallmark of several subtypes of EDS. Around 50% of patients with classic EDS harbor a heterozygous mutation in *COL5A1* or *COL5A2*, encoding the α1(V)- and the α2(V)-collagen chains, respectively. The prevalence of classic EDS has been estimated to be 1:20,000 [6]. The diagnosis of the classic type is established by clinical examination and family history [5].

Six main descriptive types were substituted for earlier types numbered with Roman numerals: classic type (EDS I and EDS II), hypermobility type (EDS III), vascular type (EDS IV), kyphoscoliosis type (EDS VI), arthrochalasia type (EDS VIIA and VIIB), and dermatosparaxis type (EDS VIIC). The key features of classic EDS I and EDS II are joint laxity and fragile, atrophic scarring skin.

Interactions between Schwann cells and neurons within the extracellular matrix are important for many aspects of peripheral nervous system development and function. Type V collagen has been associated with adhesions to heparan sulfate in Schwann cells, inhibiting the outgrowth of axons from rat neurons and promoting Schwann cell migration.

BIOMARKERS OF TYPE V COLLAGEN

Few biomarkes for type V collagen exist, although novel markers have been developed to evaluate the role of type V collagen in fibrosis-related diseases (Column 5 of Table 1). The Pro-C5 marker is an excellent example of a serological marker developed specifically to reflect true type V collagen formation [7]. Fibril-forming collagens are synthesized as precursor molecules with propeptide extensions at the N and C termini of the molecule. The mature propeptide is cleaved from procollagen by N-terminal or C-terminal proteinases, and mature collagen is integrated into the extracellular matrix [8,9]. The removal of the propeptide may be incomplete, leaving the propeptide attached to the molecule, thereby resulting in thin fibrils with abnormal crosslinks and susceptibility to rapid metabolic turnover [10,11]. Thus an internal procollagen epitope can be a marker of both formation and degradation. Pro-C5 solely measures type V collagen formation as that epitope is released when the propeptide is cleaved from the main protein during protein synthesis and incorporation into the matrix. The measured epitope is generated during this protease cleavage and is consequently an N-positioned neoepitope of the native cleaved collagen [12]. Another novel marker for type V collagen is a marker used to evaluate the matrix metalloproteinase degradation of type V collagen by using a monoclonal antibody specific toward a type V collagen fragment released into circulation in patients and in preclinical models [13]. Both the Pro-C5 and C5M markers have been shown to be correlated to portal hypertension in patients with chronic liver disease and have proven useful in preclinical models of liver disease [12,14,15]. C5M is also elevated in patients with ankylosing spondylitis compared with healthy controls [13] and decreased in patients with a hernia compared with controls [16].

Table 1 Type V Collagen

Collagen	**Description**	**References**
Gene ID (NCBI)	1289 (α1), 1290 (α2), 50509 (α3)	[17] [18]
Collagen superfamily mutations with diseases in humans	Fibrillar collagen family Ehlers–Danlos syndrome type 1: COL5A1, gene MIM number 120215; COL5A2, gene MIM number 120190 Ehlers–Danlos syndrome type 2: COL5A1, gene MIM number 120215; COL5A2, gene MIM number 120190	[19–23]
Null mutation in mice	Col5 α1 mutation mice synthesize and secrete normal levels of type I collagen, but collagen fibrils are absent and mice die at the embryonic stage Col5 α2 mutation mice survive poorly, maybe due to complications from spinal deformities, and exhibit skin and eye abnormalities	[24,25]
Tissue distribution in healthy states	1. colon, endometrium, skeletal muscle, lymph node, tonsil, thyroid gland, lung, cornea, bone, fetal membranes 2. N.A. 3. small intestines, testis, cervix, tonsil, parathyroid gland	Protein atlas, http://www.proteinatlas.org/ [3]
Tissue distribution in pathological affected states	Found in interstitial tissue, associated with type I, dermis, tendon, cornea, lung fibrosis	[26,27] [28]
Special domains	Signal peptide; TSPN-1 domain; VAR; COL1/COL2 domains; NC1 domain; NC2 domain; Laminin G–like domain; fibrillar collagen domain; VWFC binding; collagen-like-1, -2, -3, -4, -5, -6	Uniprot, http://www.uniprot.org/ [24]
Special neoepitopes	G439-V in α1/α2 and G455-L in α3 all by MMP-9; S254-Q and D1595-D by BMP-1; R1585-N by furin; P/A438-E by ADAMTS2 α1: T1584-R; P924-R; α2: G355-Q, G229-Q; α3:G1317-H, G525-R	[13,29–31]

Type V Collagen	Description	References
Protein structure and function	Type V collagen has three varieties of α chains: α1 (COL5A1), α2 (COL5A2), and α3 (COL5A3), forming a triple helix containing all three chains	
Binding proteins	Type V collagen typically forms heterofibrils with types I and III collagens	
Known central function	Essential for fibrillation of types I and III collagen	[3] [25]
Animals models with protein affected	Col5α1 and col5α2 absent	[24,25]
Biomarkers	Pro-C5, C5M	[7,13]

NCBI, National Center for Biotechnology Information; *MIM*, Mendelian Inheritance in Man; *N.A.*, not applicable.

REFERENCES

[1] Birk DE, Fitch JM, Babiarz JP, Doane KJ, Linsenmayer TF. Collagen fibrillogenesis in vitro: interaction of types I and V collagen regulates fibril diameter. J Cell Sci 1990;95(Pt 4):649–57.

[2] Kadler KE, Holmes DF, Trotter JA, Chapman JA. Collagen fibril formation. Biochem J 1996;316(Pt 1):1–11.

[3] Gelse K, Poschl E, Aigner T. Collagens–structure, function, and biosynthesis. Adv Drug Deliv Rev 2003;55:1531–46.

[4] Fichard A, Kleman JP, Ruggiero F. Another look at collagen V and XI molecules. Matrix Biol 1995;14:515–31.

[5] Beighton P, De PA, Steinmann B, Tsipouras P, Wenstrup RJ. Ehlers–Danlos syndromes: revised nosology, Villefranche, 1997. Ehlers–Danlos National Foundation (USA) and Ehlers–Danlos Support group (UK). Am J Med Genet 1998;77:31–7.

[6] Byers P. Disorders of collagen biosynthesis and structure. In: Scriver CR, Beaudet AR, Sly WS, Valle D, editors. The metabolic and molecular bases of inherited disease. Edinburgh: Churchill Livingstone; 2001. p. 1065–81.

[7] Vassiliadis E, Veidal SS, Simonsen H, Larsen DV, Vainer B, Chen X, et al. Immunological detection of the type V collagen propeptide fragment, PVCP-1230, in connective tissue remodeling associated with liver fibrosis. Biomarkers 2011;16:426–33.

[8] Gelse K, Poschl E, Aigner T. Collagens–structure, function, and biosynthesis. Adv Drug Deliv Rev 2003;55:1531–46.

[9] Schuppan D. Structure of the extracellular matrix in normal and fibrotic liver: collagens and glycoproteins. Semin Liver Dis 1990;10:1–10.

[10] Niemela O, Risteli L, Parkkinen J, Risteli J. Purification and characterization of the N-terminal propeptide of human type III procollagen. Biochem J 1985;232:145–50.

[11] Wang WM, Ge G, Lim NH, Nagase H, Greenspan DS. TIMP-3 inhibits the procollagen N-proteinase ADAMTS-2. Biochem J 2006;398:515–9.

[12] Leeming DJ, Veidal SS, Karsdal MA, Nielsen MJ, Trebicka J, Busk T, et al. Pro-C5, a marker of true type V collagen formation and fibrillation, correlates with portal hypertension in patients with alcoholic cirrhosis. Scand J Gastroenterol 2015;50:584–92.

[13] Veidal SS, Larsen DV, Chen X, Sun S, Zheng Q, Bay-Jensen AC, Leeming DJ, Nawrocki A, Larsen MR, Schett G, et al. MMP mediated type V collagen degradation (C5M) is elevated in ankylosing spondylitis. Clin Biochem 2012;45:541–6.

[14] Leeming DJ, Karsdal MA, Byrjalsen I, Bendtsen F, Trebicka J, Nielsen MJ, et al. Novel serological neo-epitope markers of extracellular matrix proteins for the detection of portal hypertension. Aliment Pharmacol Ther 2013;38:1086–96.

[15] Leeming DJ, Byrjalsen I, Jimenez W, Christiansen C, Karsdal MA. Protein fingerprinting of the extracellular matrix remodelling in a rat model of liver fibrosis – a serological evaluation. Liver Int 2013;33:439–47.

[16] Henriksen NA, Mortensen JH, Sorensen LT, Bay-Jensen AC, Agren MS, Jorgensen LN, et al. The collagen turnover profile is altered in patients with inguinal and incisional hernia. Surgery 2015;157:312–21.

[17] Greenspan DS, Byers MG, Eddy RL, Cheng W, Jani-Sait S, Shows TB. Human collagen gene COL5A1 maps to the q34.2–q34.3 region of chromosome 9, near the locus for nail-patella syndrome. Genomics 1992;12:836–7.

[18] Imamura Y, Scott IC, Greenspan DS. The pro-alpha3(V) collagen chain. Complete primary structure, expression domains in adult and developing tissues, and comparison to the structures and expression domains of the other types V and XI procollagen chains. J Biol Chem 2000;275:8749–59.

[19] Zoppi N, Gardella R, De PA, Barlati S, Colombi M. Human fibroblasts with mutations in COL5A1 and COL3A1 genes do not organize collagens and fibronectin in the extracellular matrix, down-regulate alpha2beta1 integrin, and recruit alphavbeta3 Instead of alpha5beta1 integrin. J Biol Chem 2004;279:18157–68.

[20] Bouma P, Cabral WA, Cole WG, Marini JC. COL5A1 exon 14 splice acceptor mutation causes a functional null allele, haploinsufficiency of alpha 1(V) and abnormal heterotypic interstitial fibrils in Ehlers–Danlos syndrome II. J Biol Chem 2001;276:13356–64.

[21] Michalickova K, Susic M, Willing MC, Wenstrup RJ, Cole WG. Mutations of the alpha2(V) chain of type V collagen impair matrix assembly and produce Ehlers–Danlos syndrome type I. Hum Mol Genet 1998;7:249–55.

[22] Richards AJ, Martin S, Nicholls AC, Harrison JB, Pope FM, Burrows NP. A single base mutation in COL5A2 causes Ehlers–Danlos syndrome type II. J Med Genet 1998;35:846–8.

[23] Nicholls AC, Oliver JE, McCarron S, Harrison JB, Greenspan DS, Pope FM. An exon skipping mutation of a type V collagen gene (COL5A1) in Ehlers–Danlos syndrome. J Med Genet 1996;33:940–6.

[24] Symoens S, Malfait F, Vlummens P, Hermanns-Le T, Syx D, De PA. A novel splice variant in the N-propeptide of COL5A1 causes an EDS phenotype with severe kyphoscoliosis and eye involvement. PLoS One 2011;6:e20121.

[25] Andrikopoulos K, Liu X, Keene DR, Jaenisch R, Ramirez F. Targeted mutation in the col5a2 gene reveals a regulatory role for type V collagen during matrix assembly. Nat Genet 1995;9:31–6.

[26] Martin P, Teodoro WR, Velosa AP, de MJ, Carrasco S, Christmann RB, Goldenstein-Schainberg C, Parra ER, Katayama ML, Sotto MN, et al. Abnormal collagen V deposition in dermis correlates with skin thickening and disease activity in systemic sclerosis. Autoimmun Rev 2012;11:827–35.

[27] Vittal R, Mickler EA, Fisher AJ, Zhang C, Rothhaar K, Gu H, Brown KM, Emtiazdjoo A, Lott JM, Frye SB, et al. Type V collagen induced tolerance suppresses collagen deposition, TGF-beta and associated transcripts in pulmonary fibrosis. PLoS One 2013;8:e76451.

[28] Parra ER, Teodoro WR, Velosa AP, de Oliveira CC, Yoshinari NH, Capelozzi VL. Interstitial and vascular type V collagen morphologic disorganization in usual interstitial pneumonia. J Histochem Cytochem 2006;54:1315–25.
[29] Niyibizi C, Chan R, Wu JJ, Eyre D. A 92 kDa gelatinase (MMP-9) cleavage site in native type V collagen. Biochem Biophys Res Commun 1994;202:328–33.
[30] Bonod-Bidaud C, Beraud M, Vaganay E, Delacoux F, Font B, Hulmes DJ, et al. Enzymatic cleavage specificity of the proalpha1(V) chain processing analysed by site-directed mutagenesis. Biochem J 2007;405:299–306.
[31] Colige A, Ruggiero F, Vandenberghe I, Dubail J, Kesteloot F, Van BJ, et al. Domains and maturation processes that regulate the activity of ADAMTS-2, a metalloproteinase cleaving the aminopropeptide of fibrillar procollagens types I-III and V. J Biol Chem 2005;280:34397–408.

Chapter 6

Type VI Collagen

S. Sun, M.A. Karsdal
Nordic Bioscience, Herlev, Denmark

Chapter Outline

SUMMARY
Type VI collagen is a unique beaded filament collagen. It is found in the interface between the basement membrane and interstitial matrix, where it forms a unique microfibrillar network. Type VI collagen has many binding partners. Mutations in type VI collagen have, in particular, been associated with muscle weakness disorders such as Bethlem myopathy and other muscle dystrophies. In addition to the structural and binding parameters of type VI collagen, the C-terminal propeptide is a hormone called endotrophin, associated with the metabolic syndrome. Consequently, type VI collagen is both a structural and a signaling protein. Furthermore, type VI collagen may act as an early sensor of the injury/repair response and may regulate fibrogenesis by modulating cell–cell interactions; stimulate the proliferation of mesenchymal cells; and prevent cell apoptosis. Biomarkers of type VI collagen formation and degradation have been developed, which have been shown to be regulated in connective diseases.

Type VI collagen is a unique beaded filament collagen. Type VI collagen forms a unique microfibrillar network and is expressed in the extracellular matrix (ECM) of many tissues, including the dermis, skeletal muscle, lung, blood vessels, cornea, tendon, skin, cartilage, intervertebral discs, and adipose [1–4]. Type VI collagen is found near the basement membrane and has several binding sites in that membrane [5].

Type VI collagen is composed of three different chains: α1, α2, α3. Each chain contains a short collagenous region and globular regions at both the N and C terminus [6]. Both α1 and α2 chains contain one N-terminal subdomain and two C-terminal subdomains, which are highly homologous to type A domains of von Willebrand factor (vWF-A) [7]. The α3 chain is much larger than the α1 and α2 chains. It contains 10 N-terminal subdomains (N1–N10), two C-terminal vWF-A–like subdomains (C1 and C2), and three specific C-terminal subdomains (C3–C5) [8]. Type VI collagen forms polymers intracellularly before secretion to the ECM. Dimers are formed by two monomers in an antiparallel way. Subsequently, the dimers form tetramers in a side-by-side parallel mode. Then, type VI collagen is secreted into the ECM where the microfibril forms [6,9]. Several subdomains in the N and C terminus are involved in the triple helix and microfibril formation. It is reported that the C1 domain plays critical roles in chain selection and assists the triple helix formation [8,10]. The N5 domain of the α3 chain is suggested to be involved in the assembly of microfibrils [11]. The C5 domain of the α3 chain, also called endotrophin, is important for mature microfibril formation [8] and is immediately cleaved off from the mature type VI microfibril after secretion [8,12]. The in vivo protease or possible proteases induce the release of endotrophin, and the cleavage sites are still unknown. However, it is suggested that the cleavage site is between the C1 and C2 domains [8]. In one study, an increase of incorrectly folded type VI collagen was found surrounding adipocytes in matrix metalloproteinase MMP-11–deficient mice [13]. Therefore, MMP-11 is suggested to be involved in the C5 cleavage process. However, this speculation needs to be verified.

Recently, three more chains, α4, α5, and α6, were discovered that are homologous to the α3 chain [14–16]. The α4 chain is absent in humans due to the disruption by a chromosome break in the *COL6A4* gene [16]. All three new chains are suggested to potentially substitute for the α3 chain in type VI collagen [15,17]. However, this substitution is still debated because α5 and α6 chains are also expressed in the regions where α1, α2, and α3 chains are not expressed [16].

Type VI collagen binds to a wide range of ECM proteins, including type IV collagen [5], decorin [18], biglycan [19], perlecan [20], NG2 proteoglycan [21], fibronectin [22], tenascin [23], and integrin [24]. It also associates with collagen fibrils, potentially by mediation of proteoglycans [25,26]. In addition, a series of cells including fibroblasts, chondrocytes, hematopoietic cells, and tumor cells can bind to type VI collagen through the cell binding sequence Arg-Gly-Asp in the triple helix domain [27–30]. It is widely accepted that type VI collagen aids cell attachment and connects related tissues to the surrounding matrix [3,31]. Except for the connection functions, type VI collagen may act as an early sensor of the injury/repair response and may regulate fibrogenesis by modulating cell–cell interactions. It could stimulate the proliferation of mesenchymal cells and prevent cell apoptosis [32,33]. The persistent activation of mesenchymal cells into myofibroblasts could eventually induce ECM deposition and tissue fibrosis [34]. In lung fibrosis, type VI collagen is found to be increasingly expressed [35]. The newly found α6 chain is upregulated in fibrotic areas of Duchenne muscular dystrophy muscle [36]. In addition, endotrophin can trigger adipose tissue fibrosis [37]. It also contributes to inflammation and angiogenesis, thereby facilitating tumor growth and progression [38].

Mutations COL6A1, COL6A2, and COL6A3 usually cause muscle weakness disorders such as Bethlem myopathy, Ullrich congenital muscular dystrophy, limb-girdle muscular dystrophy phenotype, and autosomal recessive myosclerosis [39–41]. Type VI collagen functions have been further investigated in knockout mice. Triple helix type VI collagen is absent in Col6a1 knockout mice. Fiber necrosis and pronounced variations in fiber diameter are found [42], which demonstrates that type VI collagen helps maintain the integrity of skeletal muscle. In addition to muscular damage, Col6a1 knockout mice also show accelerated development of osteoarthritis [43]. In another study, type VI collagen knockout has protective roles after myocardial infarction, potentially by limiting chronic apoptosis, fibrosis, and scarring [44].

BIOMARKERS OF TYPE VI COLLAGEN

Due to the specific roles of type VI collagen in the ECM environment, type VI collagen fragments generated during degradation or formation could be used as potential biomarkers for different disorders. Degradation fragment of type VI collagen induced by MMP-2 and MMP-9 (C6M) is highly increased during liver fibrosis [45] and exacerbations of chronic obstructive pulmonary disease (COPD) [46]. Pro-C6 (type VI collagen fragment generated during formation) is decreased during exacerbations of COPD [46]. In a human immobilization–remobilization study, C6M levels after 2 weeks of immobilization were a negative biomarker for muscle hypertrophy during retraining [47]. In 56-day bed rest study, Pro-C6 is related to the anabolic and catabolic responses to unloading and reloading [48].

Type VI Collagen	Description	References
Gene name and number	COL6A1, location 21q22.3 (gene ID 1291)	NCBI
	COL6A2, location 21q22.3 (gene ID 1292)	NCBI
	COL6A3, location 2q37 (gene ID 1293)	NCBI
	COL6A4 (pseudogenes due to a chromosome break)	
	COL6A5, location 3q22.1 (gene ID 256076)	NCBI
	COL6A6, location 3q22.1 (gene ID 131873)	NCBI
Mutations with diseases in humans	Bethlem myopathy (BM) Ullrich congenital muscular dystrophy (UCMD) Limb-girdle muscular dystrophy phenotype Autosomal recessive myosclerosis	[39–41]
Null mutation in mice	In Col6A1 knockout mouse, triple helix–type VI collagen was absent. Fiber necrosis, pronounced variations in fiber diameter, and accelerated development of OA are found. However, type VI collagen knockout has protective roles after myocardial infarction.	[42–44]
Tissue distribution in healthy states	Dermis, skeletal muscle, lung, blood vessels, cornea, tendon, skin, cartilage, intervertebral discs, adipose	[1–4]
Tissue distribution in pathologically affected states	Upregulated in tissue fibrosis	[35,36]

Type VI Collagen	Description	References
Special domains	Soluble cleaved C5 domain, also called endotrophin	[8,12,37,38]
Special neoepitopes	The cleavage site between the C1 and C2 domain of the α3 chain can release a soluble C-terminal fragment, endotrophin. However, the cleavage site is still unknown.	[8]
Protein structure and function	Type VI collagen is composed of three different chains: α1, α2, and α3. The protein contains a short collagenous region and globular regions at both the N and C termini. Dimers are formed by two monomers in antiparallel manner, whereas tetramers are formed in a parallel manner. After secretion into the ECM, tetramers are assembled into a microfibrillar network. Type VI collagen is expressed in many tissues and connects related tissues to the matrix.	[1,3,6]
Binding proteins	Type IV collagen, biglycan, decorin, perlecan, NG2 proteoglycan, fibronectin, tenascin, integrin.	[5,18,21–24]
Known central function	Helps cell attachment and connects related tissues to the surrounding matrix	[3,5,31]
Animal models with protein affected	COl6A1 knockout mouse model	[42]
Biomarkers of that protein	Pro-C6, C6M	[45–48]

NCBI, National Center for Biotechnology Information; *OA*, osteoarthritis; *ECM*, extracellular matrix.

REFERENCES

[1] Gelse K, Poschl E, Aigner T. Collagens–structure, function, and biosynthesis. Adv Drug Deliv Rev 2003;55:1531–46.

[2] Iyengar P, Espina V, Williams TW, Lin Y, Berry D, Jelicks LA, et al. Adipocyte-derived collagen VI affects early mammary tumor progression in vivo, demonstrating a critical interaction in the tumor/stroma microenvironment. J Clin Invest 2005;115:1163–76.

[3] Keene DR, Engvall E, Glanville RW. Ultrastructure of type VI collagen in human skin and cartilage suggests an anchoring function for this filamentous network. J Cell Biol 1988;107: 1995–2006.

[4] Zou Y, Zhang RZ, Sabatelli P, Chu ML, Bonnemann CG. Muscle interstitial fibroblasts are the main source of collagen VI synthesis in skeletal muscle: implications for congenital muscular dystrophy types Ullrich and Bethlem. J Neuropathol Exp Neurol 2008;67: 144–54.

[5] Kuo HJ, Maslen CL, Keene DR, Glanville RW. Type VI collagen anchors endothelial basement membranes by interacting with type IV collagen. J Biol Chem 1997;272:26522–9.

[6] Baldock C, Sherratt MJ, Shuttleworth CA, Kielty CM. The supramolecular organization of collagen VI microfibrils. J Mol Biol 2003;330:297–307.

[7] Chu ML, Pan TC, Conway D, Kuo HJ, Glanville RW, Timpl R, et al. Sequence analysis of alpha 1(VI) and alpha 2(VI) chains of human type VI collagen reveals internal triplication of globular domains similar to the A domains of von Willebrand factor and two alpha 2(VI) chain variants that differ in the carboxy terminus. EMBO J 1989;8:1939–46.

[8] Lamande SR, Morgelin M, Adams NE, Selan C, Allen JM. The C5 domain of the collagen VI alpha3(VI) chain is critical for extracellular microfibril formation and is present in the extracellular matrix of cultured cells. J Biol Chem 2006;281:16607–14.
[9] Furthmayr H, Wiedemann H, Timpl R, Odermatt E, Engel J. Electron-microscopical approach to a structural model of intima collagen. Biochem J 1983;211:303–11.
[10] Ball SG, Baldock C, Kielty CM, Shuttleworth CA. The role of the C1 and C2 a-domains in type VI collagen assembly. J Biol Chem 2001;276:7422–30.
[11] Fitzgerald J, Morgelin M, Selan C, Wiberg C, Keene DR, Lamande SR, et al. The N-terminal N5 subdomain of the alpha 3(VI) chain is important for collagen VI microfibril formation. J Biol Chem 2001;276:187–93.
[12] Aigner T, Hambach L, Soder S, Schlotzer-Schrehardt U, Poschl E. The C5 domain of Col6A3 is cleaved off from the Col6 fibrils immediately after secretion. Biochem Biophys Res Commun 2002;290:743–8.
[13] Motrescu ER, Blaise S, Etique N, Messaddeq N, Chenard MP, Stoll I, et al. Matrix metalloproteinase-11/stromelysin-3 exhibits collagenolytic function against collagen VI under normal and malignant conditions. Oncogene 2008;27:6347–55.
[14] Fitzgerald J, Rich C, Zhou FH, Hansen U. Three novel collagen VI chains, alpha4(VI), alpha5(VI), and alpha6(VI). J Biol Chem 2008;283:20170–80.
[15] Gara SK, Grumati P, Urciuolo A, Bonaldo P, Kobbe B, Koch M, et al. Three novel collagen VI chains with high homology to the alpha3 chain. J Biol Chem 2008;283:10658–70.
[16] Fitzgerald J, Holden P, Hansen U. The expanded collagen VI family: new chains and new questions. Connect Tissue Res 2013;54:345–50.
[17] Sabatelli P, Gara SK, Grumati P, Urciuolo A, Gualandi F, Curci R, et al. Expression of the collagen VI alpha5 and alpha6 chains in normal human skin and in skin of patients with collagen VI-related myopathies. J Invest Dermatol 2011;131:99–107.
[18] Bidanset DJ, Guidry C, Rosenberg LC, Choi HU, Timpl R, Hook M. Binding of the proteoglycan decorin to collagen type VI. J Biol Chem 1992;267:5250–6.
[19] Wiberg C, Hedbom E, Khairullina A, Lamande SR, Oldberg A, Timpl R, et al. Biglycan and decorin bind close to the n-terminal region of the collagen VI triple helix. J Biol Chem 2001;276:18947–52.
[20] Tillet E, Wiedemann H, Golbik R, Pan TC, Zhang RZ, Mann K, et al. Recombinant expression and structural and binding properties of alpha 1(VI) and alpha 2(VI) chains of human collagen type VI. Eur J Biochem 1994;221:177–85.
[21] Stallcup WB, Dahlin K, Healy P. Interaction of the NG2 chondroitin sulfate proteoglycan with type VI collagen. J Cell Biol 1990;111:3177–88.
[22] Carter WG. Transformation-dependent alterations in glycoproteins of extracellular matrix of human fibroblasts. Characterization of GP250 and the collagen-like GP140. J Biol Chem 1982;257:13805–15.
[23] Minamitani T, Ariga H, Matsumoto K. Deficiency of tenascin-X causes a decrease in the level of expression of type VI collagen. Exp Cell Res 2004;297:49–60.
[24] Pfaff M, Aumailley M, Specks U, Knolle J, Zerwes HG, Timpl R. Integrin and Arg-Gly-Asp dependence of cell adhesion to the native and unfolded triple helix of collagen type VI. Exp Cell Res 1993;206:167–76.
[25] Keene DR, Ridgway CC, Iozzo RV. Type VI microfilaments interact with a specific region of banded collagen fibrils in skin. J Histochem Cytochem 1998;46:215–20.
[26] Nakamura M, Kimura S, Kobayashi M, Hirano K, Hoshino T, Awaya S. Type VI collagen bound to collagen fibrils by chondroitin/dermatan sulfate glycosaminoglycan in mouse corneal stroma. Jpn J Ophthalmol 1997;41:71–6.

[27] Aumailley M, Mann K, von der MH, Timpl R. Cell attachment properties of collagen type VI and Arg-Gly-Asp dependent binding to its alpha 2(VI) and alpha 3(VI) chains. Exp Cell Res 1989;181:463–74.

[28] Doane KJ, Yang G, Birk DE. Corneal cell-matrix interactions: type VI collagen promotes adhesion and spreading of corneal fibroblasts. Exp Cell Res 1992;200:490–9.

[29] Klein G, Muller CA, Tillet E, Chu ML, Timpl R. Collagen type VI in the human bone marrow microenvironment: a strong cytoadhesive component. Blood 1995;86:1740–8.

[30] Han J, Daniel JC. Biosynthesis of type VI collagen by glioblastoma cells and possible function in cell invasion of three-dimensional matrices. Connect Tissue Res 1995;31:161–70.

[31] Bonaldo P, Russo V, Bucciotti F, Doliana R, Colombatti A. Structural and functional features of the alpha 3 chain indicate a bridging role for chicken collagen VI in connective tissues. Biochemistry 1990;29:1245–54.

[32] Ruhl M, Johannsen M, Atkinson J, Manski D, Sahin E, Somasundaram R, et al. Soluble collagen VI induces tyrosine phosphorylation of paxillin and focal adhesion kinase and activates the MAP kinase erk2 in fibroblasts. Exp Cell Res 1999;250:548–57.

[33] Ruhl M, Sahin E, Johannsen M, Somasundaram R, Manski D, Riecken EO, et al. Soluble collagen VI drives serum-starved fibroblasts through S phase and prevents apoptosis via down-regulation of Bax. J Biol Chem 1999;274:34361–8.

[34] Hutchison N, Fligny C, Duffield JS. Resident mesenchymal cells and fibrosis. Biochim Biophys Acta 2013;1832:962–71.

[35] Specks U, Nerlich A, Colby TV, Wiest I, Timpl R. Increased expression of type VI collagen in lung fibrosis. Am J Respir Crit Care Med 1995;151:1956–64.

[36] Sabatelli P, Gualandi F, Gara SK, Grumati P, Zamparelli A, Martoni E, et al. Expression of collagen VI alpha5 and alpha6 chains in human muscle and in Duchenne muscular dystrophy-related muscle fibrosis. Matrix Biol 2012;31:187–96.

[37] Sun K, Park J, Gupta OT, Holland WL, Auerbach P, Zhang N, et al. Endotrophin triggers adipose tissue fibrosis and metabolic dysfunction. Nat Commun 2014;5:3485.

[38] Chen P, Cescon M, Bonaldo P. Collagen VI in cancer and its biological mechanisms. Trends Mol Med 2013;19:410–7.

[39] Bushby KM, Collins J, Hicks D. Collagen type VI myopathies. Adv Exp Med Biol 2014;802:185–99.

[40] Lampe AK, Bushby KM. Collagen VI related muscle disorders. J Med Genet 2005;42:673–85.

[41] Lampe AK, Zou Y, Sudano D, O'Brien KK, Hicks D, Laval SH, et al. Exon skipping mutations in collagen VI are common and are predictive for severity and inheritance. Hum Mutat 2008;29:809–22.

[42] Bonaldo P, Braghetta P, Zanetti M, Piccolo S, Volpin D, Bressan GM. Collagen VI deficiency induces early onset myopathy in the mouse: an animal model for Bethlem myopathy. Hum Mol Genet 1998;7:2135–40.

[43] Alexopoulos LG, Youn I, Bonaldo P, Guilak F. Developmental and osteoarthritic changes in Col6a1-knockout mice: biomechanics of type VI collagen in the cartilage pericellular matrix. Arthritis Rheum 2009;60:771–9.

[44] Luther DJ, Thodeti CK, Shamhart PE, Adapala RK, Hodnichak C, Weihrauch D, et al. Absence of type VI collagen paradoxically improves cardiac function, structure, and remodeling after myocardial infarction. Circ Res 2012;110:851–6.

[45] Veidal SS, Karsdal MA, Vassiliadis E, Nawrocki A, Larsen MR, Nguyen QH, et al. MMP mediated degradation of type VI collagen is highly associated with liver fibrosis–identification and validation of a novel biochemical marker assay. PLoS One 2011;6:e24753.

[46] Sand JM, Knox AJ, Lange P, Sun S, Kristensen JH, Leeming DJ, et al. Accelerated extracellular matrix turnover during exacerbations of COPD. Respir Res 2015;16:69.
[47] Nedergaard A, Sun S, Karsdal MA, Henriksen K, Kjaer M, Lou Y, et al. Type VI collagen turnover-related peptides-novel serological biomarkers of muscle mass and anabolic response to loading in young men. J Cachexia Sarcopenia Muscle 2013;4:267–75.
[48] Sun S, Henriksen K, Karsdal MA, Byrjalsen I, Rittweger J, Armbrecht G, et al. Collagen type III and VI turnover in response to long-term immobilization. PLoS One 2015;10:e0144525.

Chapter 7

Type VII Collagen

J.H. Mortensen, M.A. Karsdal
Nordic Bioscience, Herlev, Denmark

Chapter Outline

SUMMARY

Type VII collagen (COLVII) is primarily synthesized by keratinocytes and fibroblasts. It is crucial for the function and stability of the extracellular matrix (ECM) as it is an anchoring fibril collagen, where it binds type I and type III collagen, providing stability to the interstitial membrane and the basement membrane ECM structures. The NC1 can be subdivided into smaller components comprising adhesive proteins with high-binding affinity to basement membrane proteins, especially laminin-5 and laminin-6 and type IV collagen. The main mutation investigated for COLVII is *epidermolysis bullosa*, a severe skin disorder. COLVII has also been reported to be involved in autoimmune diseases such as systemic lupus erythematosus, Sjögren syndrome, and systemic sclerosis. The function and roles of COLVII are emerging, and no biomarkers are currently developed.

Type VII collagen (COLVII) is primarily synthesized by keratinocytes and fibroblasts. COLVII is approximately 424 nm long [1]. The designated gene, named COL7A1, is located on chromosome 3. COLVII is strongly associated with the epithelia segment of tissues, especially the dermis. However, COLVII has also been reported to be present in other tissues including those of the rectum, colon, small intestines, esophagus, oral mucosa, cervix, placenta, skeletal muscle, and cornea [1,2].

COLVII is crucial for the function and stability of the extracellular matrix (ECM) as it is an anchoring fibril collagen [1]. COLVII consists of three components, the N-terminal noncollagenous-1 (NC1) part (~145 kDa), the triple α-chain helix collagenous domain, and the C-terminal noncollagenous-2 (NC2) part (~30 kDa) (Fig. 3d, Introduction) [3–6]. NC1 is the major component of COLVII and determines the ECM function due to its role as an anchoring fibril [1]. The NC1 can be subdivided into smaller components comprising adhesive proteins with high binding affinity to basement membrane proteins, especially laminin-5 and -6 and type IV collagen. These adhesive proteins are mainly comprised of segments with homology with von Willebrand factor, cartilage matrix proteins, and fibronectin type III domains [7,8].

For COLVII to be a fully functional anchoring fibril, two COLVII will attach at the C-terminal end to form a U-shaped structure (Fig. 4, Introduction) [9]. The U-shaped structure is now a COLVII dimer that functions as a trap for the interstitial collagen fibrils in the triple helix of COLVII. For the NC2 domains to assemble and form the U-shaped COLVII, the NC2 domain is removed by proteases. Moreover, the NC2 domain also includes a segment with homology to the Kunitz protease inhibitor [10], thus stopping the proteolytic cleavage of the C terminus. The interstitial type I, III, and V collagens are the major targets of entrapment by the U-shaped COLVII. In addition, the NC1 domain will bind basement membrane proteins, especially laminins and type IV collagens, and together with the U-shape formation facilitated by the NC2 domain, COLVII provide stability to the interstitial membrane and the basement membrane ECM structures [7,8].

COLVII mutations that lead to defects and loss of its function have critical impacts on the integrity of the ECM and thus of the tissue [11,12]. The skin disorder dystrophic epidermolysis bullosa (DEB) is directly related to mutations in the COL7A1 gene [11], whereas in epidermolysis bullosa acquisita (EBA), autoantibodies are generated against COLVII [2,9]. Hence, COLVII is also referred to as the *epidermolysis bullosa acquisita-antigen* [2]. Mutations affecting both alleles of the COL7A1 gene result in the absence of the COLVII protein [13]. The disease manifestations are similar in both DEB and EBA, which are primarily blistering and scarring of the skin [2,11].

EBA has also been linked to inflammatory bowel disease, especially Crohn disease (CD). In 40% of all CD patients autoantibodies generated against COLVII have been found to be present in the serum [2]. Furthermore, CD can evolve so that EBA will manifest in some patients. Circulating autoantibodies against COLVII have been linked to other autoimmune diseases including systemic lupus erythematosus and Sjögren syndrome [14]. Recessive dystrophic epidermolysis bullosa (RDEB), is a type of DEB, but lacks collagen type VII epitopes. Interestingly RDEB has also been associated with aggressive squamous cell carcinomas. These cancer forms usually develop early in life and have very high metastatic rates, which drastically decrease the life span of affected patients. COLVII has also been shown to be upregulated in cancer stem cells cultured as spheroids in vitro [15].

Increased expression and accumulation of COLVII in the skin of patients with systemic sclerosis (SSc), and accumulation of COLVII in SSc patients was directly regulated by transforming growth factor-β [16]. Moreover, it has been shown that matrix metalloproteinases (MMPs) can cleave COLVII [17] and generate fragments that can be used to develop neoepitope biomarker assays. COLVII neoepitope assays could potentially add valuable information in detecting and managing diseases, such as DEB, EBA, and CD, where COLVII is affected since the neoepitopes may reflect chronic MMP activity that will continuously degrade COLVII. Furthermore, these fragments could be the potential triggers for generating antibodies against COLVII.

Collagen Type VII	Description	References
Gene name and number	COL7A1	[1,12]
Mutations with diseases in humans	Dystrophic epidermolysis bullosa	[2]
Null mutation in mice	COLVII knockout mice presented with severe blistering of skin at birth, which resembles dystrophic epidermolysis bullosa in humans	[12]
Tissue distribution in healthy states	Skin, rectum, colon, small intestines, esophagus, oral mucosa, cervix, placenta, skeletal muscle, and cornea	[1,2]
Tissue distribution in pathologically affected states	Skin is affected in dystrophic epidermolysis bullosa as well as in 40% of Crohn disease patients. COLVII has also been reported to be involved the autoimmune diseases eg, systemic lupus erythematosus, Sjögren syndrome, and systemic sclerosis.	[2,14–16]
Special domains	The NC1 is located in the N terminus, in the collagenous domain containing the triple helix COLVII formation and in the C-terminal NC2 domain	[5,6]
Special neoepitopes	Interstitial MMP cleavage of COLVII	[17]
Protein structure and function	The NC1 domain has high affinity to proteins of the basement membrane, such as laminins and COLIV. The collagenous triple α-chain helix domain entraps interstitial collagens. The NC2 domain is responsible for dimer formation of COLVII, which facilitates entrapment of other collagens	[11]
Binding proteins	NC1 (von Willebrand factor)	[3–6]
Known central function	Attachment to the basement membrane via the NC1 domain, and interstitial collagen entrapment	[7]
Animals models with protein affected	The COLVII knockout mouse is a model for dystrophic epidermolysis bullosa	[12]
Biomarkers	Autoantibodies against COLVII	[18]

COL, Collagen; *MMP*, matrix metalloproteinase.

REFERENCES

[1] Sakai L. Type VII collagen is a major structural component of anchoring fibrils. J Cell Biol 1986;103:1577–86.

[2] Chen M, O'Toole E, Sanghavi J, Mahmud N, Kelleher D, Weir D, et al. The epidermolysis bullosa acquisita antigen (type VII collagen) is present in human colon and patients with Crohn's disease have autoantibodies to type VII collagen. J Invest Dermatol 2002;118:1059–64.

[3] Christiano AM, Hoffman GG, Chung-Honet LC, Lee S, Cheng W, Uitto J, et al. Structural organization of the human type VII collagen gene (COL7A1), composed of more exons than any previously characterized gene. Genomics 1994;21:169–79.

[4] Christiano AM. Cloning of human type VII collagen: complete primary sequence of the alpha-1(VII) chain and identification of intragenic polymorphisms. J Biol Chem 1994;269:20256–62.

[5] Christiano AM. The large non-collagenous domain (NC-1) of type VII collagen is amino-terminal and chimeric: homology to cartilage matrix protein, the type III domains of fibronectin and the A domains of von Willebrand factor. Hum Mol Genet 1992;1:475–81.

[6] Greenspan DS. The carboxyl-terminal half of type VII collagen, including the noncollagenous NC-2 domain and intron/exon organization of the corresponding region of the COL7A1 gene. Hum Mol Genet 1993;2:273–8.

[7] Brittingham R, Uitto J, Fertala A. High-affinity binding of the NC1 domain of collagen VII to laminin 5 and collagen IV. Biochem Biophys Res Commun 2006;343:692–9.

[8] Villone D, Fritsch A, Koch M, Bruckner-Tuderman L, Hansen U, Bruckner P. Supramolecular interactions in the dermo-epidermal junction zone: anchoring fibril-collagen VII tightly binds to banded collagen fibrils. J Biol Chem 2008;283:24506–13.

[9] Shimizu H. Most anchoring fibrils in human skin originate and terminate in the lamina densa. Lab Invest 1997;76:753–63.

[10] Colombo M, Brittingham RJ, Klement JF, Majsterek I, Birk DE, Uitto J, et al. Procollagen VII self-assembly depends on site-specific interactions and is promoted by cleavage of the NC2 domain with procollagen C-proteinase. Biochemistry 2003;42:11434–42.

[11] Bruckner-Tuderman L, Mitsuhashi Y, Schnyder UW, Bruckner P. Anchoring fibrils and type VII collagen are absent from skin in severe recessive dystrophic epidermolysis bullosa. J Invest Dermatol 1989;93:3–9.

[12] Heinonen S. Targeted inactivation of the type VII collagen gene (Col7a1) in mice results in severe blistering phenotype: a model for recessive dystrophic epidermolysis bullosa. J Cell Sci 1999;112(Pt 2):3641–8.

[13] Uitto J, Richard G. Progress in epidermolysis bullosa: from eponyms to molecular genetic classification. Clin Dermatol 2005;23:33–40.

[14] Fujii K. Detection of anti-type VII collagen antibody in Sjögren's syndrome/lupus erythematosus overlap syndrome with transient bullous systemic lupus erythematosus. Br J Dermatol 1998;139:302–6.

[15] Oktem G, Sercan O, Guven U, Uslu R, Uysal A, Goksel G, et al. Cancer stem cell differentiation: TGFβ1 and versican may trigger molecules for the organization of tumor spheroids. Oncol Rep 2014;32:641–9.

[16] Rudnicka L, Varga J, Christiano AM, Iozzo RV, Jimenez SA, Uitto J. Elevated expression of type VII collagen in the skin of patients with systemic sclerosis. Regulation by transforming growth factor-beta. J Clin Invest 1994;93:1709–15.

[17] Seltzer JL. Cleavage of type VII collagen by interstitial collagenase and type IV collagenase (gelatinase) derived from human skin. J Biol Chem 1989;264:3822–6.

[18] Woodley DT, Cogan J, Wang X, Hou Y, Haghighian C, Kudo G, et al. De novo anti-type VII collagen antibodies in patients with recessive dystrophic epidermolysis bullosa. J Invest Dermatol 2014;134:1138–40.

Chapter 8

Type VIII Collagen

N.U.B. Hansen, M.A. Karsdal
Nordic Bioscience, Herlev, Denmark

Chapter Outline

SUMMARY
Type VIII collagen is a short-chain, nonfibrillar collagen and the major component of the Descemet's membrane. Type VIII collagen is produced by endothelial cells and forms unique hexagonal lattice structures. It is found in the heart, brain, liver, lung, and muscles and around chondrocytes in cartilage. Furthermore, type VIII collagen is found around actively proliferating vessels of brain tumors and in large fibrotic vessels of angiomas. Mutations are involved in Fuchs endothelial dystrophy. Type VIII collagen both holds structural and signaling properties, as exemplified by vastatins potent inhibitory effects on angiogensis. There are currently both N- and C-terminal antibodies incorporated in enzyme-linked immunosorbent assays for type VIII collagen.

Type VIII collagen is a short-chain, nonfibrillar collagen. It is the major component of the Descemet membrane—the basement membrane separating corneal endothelial cells from corneal stroma—of corneal endothelial cells. It is part of the endothelium of blood vessels and is present in arterioles and venules; thus, it is found in the heart, brain, liver, and lung and around chondrocytes in cartilage [1]. The human α1 procollagen gene is located on chromosome 3, whereas the human α2 procollagen gene is located on chromosome 1. Each α chain has a molecular mass of approximately 60 kDa [2]. Previously, type VIII collagen was described as a heterotrimer comprised of two α1 chains and one α2 chain [3], but *in vitro* studies have shown that homotrimers of either α1 or α2 are formed [4]. These homotrimers are pepsin resistant and an immunohistochemistry study showed they did not always colocalize in the cornea, optic nerve, aorta, and umbilical cord [5].

The α1 chain is 744 amino acids (aa) long, with the first 27 aa in the N-terminal being a signal peptide. The NC1 domain of type VIII collagen (aa 572–744) is

Biochemistry of Collagens, Laminins and Elastin.

a protein named vastatin, which has been shown to be a potent angiogenesis inhibitor with apoptotic-inducing activity on aortic endothelial cells [6]. The α2 chain is 703 aa long, with a signal peptide in the N-terminal at aa 1–28. The α1 and the α2 procollagen genes each contain four exons. One of the characteristic features is the largest exon encodes the entire triple helical (COL1) and the C-terminal nontriple helical (NC1) domains. Type VIII collagen shares a high sequence homology with type X collagen along with a similar intron–exon structure. This suggests that both collagens arose from the same precursor gene and belong to the same collagen subclass [7–9]. Type VIII collagen is able to form hexagonal lattice structures (Fig. 4, Introduction) like type X collagen.

Sequence analysis has shown that each α chain contains a collagenous triple helical domain of 454 aa, a long NC1 domain, and a shorter NC2 domain. Type VIII collagen is a glycoprotein that is very susceptible to neutrophil elastase and is completely degraded within 4 h, unlike other vascular collagens such as type I, II, IV and V collagen [10]. When incubated with pepsin, it will form two pepsin-resistant fragments, one fragment from each chain, of around 50 kDa [11].

Type VIII collagen is synthesized by aortic and corneal endothelial cells, pulmonary artery endothelial cells, and microvascular endothelial cells. Not all endothelial cells express type VIII collagen, since it is absent from large and small vessels [12]. In addition, it has been found that monocytes and macrophages express type VIII collagen *in vitro* and *in vivo* [13]. Human mast cells have also been shown to produce type VIII collagen under normal and pathological conditions. It has been shown to contribute to angiogenesis, tissue remodeling, fibrosis, and cancer [14–18]. Mast cells expressing type VIII collagen are found in perivascular spaces [19]. Type VIII collagen is also expressed by smooth muscle cells where it stimulates cell migration [20].

Corneal endothelium secretes type VIII collagen, which is assembled into a hexagonal lattice structure in the Descemet membrane by the interactions of α1 and α2 polypeptides. Mutations in the *COL8A2* gene are associated with early Fuchs endothelial corneal dystrophy, but there is no correlation with *COL8A1* [21]. It is believed that type VIII collagen is involved in endothelial differentiation and organization [22]. In cardiac development type VIII collagen plays an important role in vascularization where it has been immunolocalized to the subendothelium of capillaries and small arteries [23,24]. Type VIII collagen is upregulated in early atherogenesis and is suggested to be involved in thrombosis and monocyte infiltration. It is accumulated in atherosclerotic lesions, and its distribution patterns imply a role in plaque stabilization. Type VIII collagen is found around actively proliferating vessels of brain tumors and in large fibrotic vessels of angiomas [25]. Lastly, type VIII collagen is expressed in human diabetic nephropathy, but not in other renal diseases [26].

BIOMARKERS OF TYPE VIII COLLAGEN

Commercially available antibodies for the N- and C-terminal of type VIII collagen exist, but often the epitope is unspecified. The majority of the antibodies are polyclonal; hence, they have a higher cross-reactivity toward nonspecific antigens. The available monoclonal antibodies are, in general, specific for a large aa sequence. Many diseases such as fibrosis and cancer have a high tissue turnover, ie, there is an increased formation and degradation of proteins, and possibly type VIII collagen. It is highly plausible that type VIII collagen is cleaved by collagenases during tissue remodeling, and a monoclonal antibody specific for a special protease–generated fragment, or type VIII collagen telopeptides would assist in the discovery and quantification of the role of type VIII collagen in health and disease.

Type VIII Collagen	Description	References
Gene name and number	COL8A1 (3q11.1-q13.2), COL8A2 (1p34.2-p32.3)	[27,28]
Mutations with diseases in humans	Fuchs corneal dystrophy	[29]
Null mutation in mice	COl8a1 KO and COl8a2 KO develop normally and have a normal life span. COl8a1$^{-/-}$/COl8a2$^{-/-}$ mice have dysgenesis of the anterior segment of the eye.	[30]
Tissue distribution in healthy states	Bone, brain, cartilage, eye, heart, kidney, liver, lung, muscle, skin, spleen, vascular tissues, ligaments and tendons, nerves	[31]
Tissue distribution in pathological affected states	Upregulated in atherosclerosis and many tumors due to actively proliferating vessels. Expressed by mast cells in fibrotic tissues and contributes to the fibrotic changes in diabetes and diabetic nephropathy. Involved in Fuchs endothelial dystrophy	[12,19,25,26,32–34]
Special domains	NC1 domain of α1 is known as vastatin, a potent angiogenesis inhibitor with an apoptotic effect on aortic endothelial cells.	[6,35]
Special neoepitopes	Cleaved by neutrophil elastase into two 50K (M_r) fragments; 50K-A (α1) and 50K-B (α2)	[36]
Protein structure and function	Heterotrimer comprised of two α1 chains and one α2 chain. Increasing evidence suggests homotrimers of each. Each chain has an NC1, NC2, and triple helical domain.	[7]
Binding proteins	α1β1-integrin and GPIa/IIa (α2β1-integrin)	[20,37]
Known central function	Vascular remodeling and maintenance of vessel wall integrity and structure	[26,38]
Animals models with protein affected	COl8 KO mice, Aca23 mice (missense mutation)	[30,39]
Biomarkers	N- and C-terminal markers	

COL, Collagen; *KO*, knockout.

REFERENCES

[1] Kittelberger R, Davis PF, Flynn DW, Greenhill NS. Distribution of type VIII collagen in tissues: an immunohistochemical study. Connect Tissue Res 1990;24:303–18.

[2] Ma ZH, Ma JH, Jia L, Zhao YF. Effect of enhanced expression of COL8A1 on lymphatic metastasis of hepatocellular carcinoma in mice. Exp Ther Med 2012;4:621–6.

[3] Jander R, Korsching E, Rauterberg J. Characteristics and in vivo occurrence of type VIII collagen. Eur J Biochem 1990;189:601–7.

[4] Illidge C, Kielty C, Shuttleworth A. The alpha1(VIII) and alpha2(VIII) chains of type VIII collagen can form stable homotrimeric molecules. J Biol Chem 1998;273:22091–5.

[5] Greenhill NS, Ruger BM, Hasan Q, Davis PF. The alpha1(VIII) and alpha2(VIII) collagen chains form two distinct homotrimeric proteins in vivo. Matrix Biol 2000;19:19–28.

[6] Xu R, Yao ZY, Xin L, Zhang Q, Li TP, Gan RB. NC1 domain of human type VIII collagen (alpha 1) inhibits bovine aortic endothelial cell proliferation and causes cell apoptosis. Biochem Biophys Res Commun 2001;289:264–8.

[7] Sutmuller M, Bruijn JA, De HE. Collagen types VIII and X, two non-fibrillar, short-chain collagens. Structure homologies, functions and involvement in pathology. Histol Histopathol 1997;12:557–66.

[8] Yamaguchi N, Mayne R, Ninomiya Y. The alpha 1 (VIII) collagen gene is homologous to the alpha 1 (X) collagen gene and contains a large exon encoding the entire triple helical and carboxyl-terminal non-triple helical domains of the alpha 1 (VIII) polypeptide. J Biol Chem 1991;266:4508–13.

[9] Illidge C, Kielty CM, Shuttleworth CA. Stability of type VIII collagen homotrimers: comparison with alpha1(X). Biochem Soc Trans 1998;26:S18.

[10] Kittelberger R, Neale TJ, Francky KT, Greenhill NS, Gibson GJ. Cleavage of type VIII collagen by human neutrophil elastase. Biochim Biophys Acta 1992;1139:295–9.

[11] Kapoor R, Sakai LY, Funk S, Roux E, Bornstein P, Sage EH. Type VIII collagen has a restricted distribution in specialized extracellular matrices. J Cell Biol 1988;107:721–30.

[12] Sage H, Balian G, Vogel AM, Bornstein P. Type VIII collagen. Synthesis by normal and malignant cells in culture. Lab Invest 1984;50:219–31.

[13] Weitkamp B, Cullen P, Plenz G, Robenek H, Rauterberg J. Human macrophages synthesize type VIII collagen in vitro and in the atherosclerotic plaque. FASEB J 1999;13:1445–57.

[14] Adiguzel E, Hou G, Mulholland D, Hopfer U, Fukai N, Olsen B, et al. Migration and growth are attenuated in vascular smooth muscle cells with type VIII collagen-null alleles. Arterioscler Thromb Vasc Biol 2006;26:56–61.

[15] Ma XJ, Dahiya S, Richardson E, Erlander M, Sgroi DC. Gene expression profiling of the tumor microenvironment during breast cancer progression. Breast Cancer Res 2009;11:R7.

[16] Wang W, Xu G, Ding CL, Zhao LJ, Zhao P, Ren H, et al. All-trans retinoic acid protects hepatocellular carcinoma cells against serum-starvation-induced cell death by upregulating collagen 8A2. FEBS J 2013;280:1308–19.

[17] Vlodavsky I, Christofori G. Fibroblast growth factors in tumor progression and angiogenesis. In: Teicher BA, editor. Antiangiogenic agents in cancer therapy. New York: Springer Science & Business Media; 1998. p. 93–119.

[18] Karttunen L, Felbor U, Vikkula M, Olsen BR. Vascular matrix and disorders. In: Zon LI, editor. Hematopoiesis: a developmental approach. New York: Oxford University Press; 2001. p. 784–95.

[19] Ruger B, Dunbar PR, Hasan Q, Sawada H, Kittelberger R, Greenhill N, et al. Human mast cells produce type VIII collagen in vivo. Int J Exp Pathol 1994;75:397–404.

[20] Adiguzel E, Hou G, Sabatini PJ, Bendeck MP. Type VIII collagen signals via beta1 integrin and RhoA to regulate MMP-2 expression and smooth muscle cell migration. Matrix Biol 2013;32:332–41.

[21] Elhalis H, Azizi B, Jurkunas UV. Fuchs endothelial corneal dystrophy. Ocul Surf 2010;8:173–84.

[22] Iruela-Arispe ML, Diglio CA, Sage EH. Modulation of extracellular matrix proteins by endothelial cells undergoing angiogenesis in vitro. Arterioscler Thromb 1991;11:805–15.

[23] Sage H, Iruela-Arispe ML. Type VIII collagen in murine development. Association with capillary formation in vitro. Ann N Y Acad Sci 1990;580:17–31.

[24] Iruela-Arispe ML, Sage EH. Expression of type VIII collagen during morphogenesis of the chicken and mouse heart. Dev Biol 1991;144:107–18.

[25] Paulus W, Sage EH, Jellinger K, Roggendorf W. Type VIII collagen in the normal and diseased human brain. Acta Histochem Suppl 1992;42:195–9.

[26] Gerth J, Cohen CD, Hopfer U, Lindenmeyer MT, Sommer M, Grone HJ, et al. Collagen type VIII expression in human diabetic nephropathy. Eur J Clin Invest 2007;37:767–73.

[27] Muragaki Y, Jacenko O, Apte S, Mattei MG, Ninomiya Y, Olsen BR. The alpha 2(VIII) collagen gene. A novel member of the short chain collagen family located on the human chromosome 1. J Biol Chem 1991;266:7721–7.

[28] Muragaki Y, Mattei MG, Yamaguchi N, Olsen BR, Ninomiya Y. The complete primary structure of the human alpha 1 (VIII) chain and assignment of its gene (COL8A1) to chromosome 3. Eur J Biochem 1991;197:615–22.

[29] Mootha VV, Gong X, Ku HC, Xing C. Association and familial segregation of CTG18.1 trinucleotide repeat expansion of TCF4 gene in Fuchs' endothelial corneal dystrophy. Invest Ophthalmol Vis Sci 2014;55:33–42.

[30] Hopfer U, Fukai N, Hopfer H, Wolf G, Joyce N, Li E, et al. Targeted disruption of Col8a1 and Col8a2 genes in mice leads to anterior segment abnormalities in the eye. FASEB J 2005;19:1232–44.

[31] Ricard-Blum S, Dublet B, van der Rest M. Unconventional collagens: types VI, VII, VIII, IX, X, XII, XIV, XVI, and XIX. Oxford University Press; 2000.

[32] Levy SG, Moss J, Sawada H, Dopping-Hepenstal PJ, McCartney AC. The composition of wide-spaced collagen in normal and diseased Descemet's membrane. Curr Eye Res 1996;15:45–52.

[33] Alitalo K, Bornstein P, Vaheri A, Sage H. Biosynthesis of an unusual collagen type by human astrocytoma cells in vitro. J Biol Chem 1983;258:2653–61.

[34] Lopes J, Adiguzel E, Gu S, Liu SL, Hou G, Heximer S, et al. Type VIII collagen mediates vessel wall remodeling after arterial injury and fibrous cap formation in atherosclerosis. Am J Pathol 2013;182:2241–53.

[35] Lauer JL, Fields GB. Collagen in Cancer. In: The tumor microenvironment. Springer-Verlag New York; 2010. p. 477–507.

[36] Kapoor R, Bornstein P, Sage EH. Type VIII collagen from bovine Descemet's membrane: structural characterization of a triple-helical domain. Biochemistry 1986;25:3930–7.

[37] Saelman EU, Nieuwenhuis HK, Hese KM, de Groot PG, Heijnen HF, Sage EH, et al. Platelet adhesion to collagen types I through VIII under conditions of stasis and flow is mediated by GPIa/IIa (alpha 2 beta 1-integrin). Blood 1994;83:1244–50.

[38] Plenz GA, Deng MC, Robenek H, Volker W. Vascular collagens: spotlight on the role of type VIII collagen in atherogenesis. Atherosclerosis 2003;166:1–11.

[39] Steinhart MR, Cone FE, Nguyen C, Nguyen TD, Pease ME, Puk O, et al. Mice with an induced mutation in collagen 8A2 develop larger eyes and are resistant to retinal ganglion cell damage in an experimental glaucoma model. Mol Vis 2012;18:1093–106.

Chapter 9

Type IX Collagen

Y. He, M.A. Karsdal
Nordic Bioscience, Herlev, Denmark

Chapter Outline

SUMMARY

Type IX collagen is a fibril-associated collagen with interrupted triple helices. Type IX collagen is present in the chondrocytes of growth-plate cartilage, adult articular, cartilage and intervertebral discs. Mutations in type IX collagen can predispose individuals to multiple epiphyseal dysplasia, a clinically highly heterogeneous skeletal disorder, with early-onset osteoarthritis as a very common manifestation. Type IX collagen contributes to the stabilization of the fibrillar collagen network in the cartilage matrix and the anchorage of matrilin 3 and proteoglycans, which controls the diameter of collagen fibrils; consequently, it is an essential part of articular cartilage. There are currently no biomarkers of type IX collagen, which is needed to further elucidate the role of type IX collagen in health and disease.

Type IX collagen is a fibril-associated collagen with interrupted triple helices. Each type IX collagen molecule consists of three chains: one α1 chain, one α2 chain, and one α3 chain; each of these chains is encoded by three genes: COL9A1 (gene ID: 1297), COL9A2 (gene ID: 1298), and COL9A3 (gene ID: 1299), respectively [1]. Studies suggest that mutations identified in COL9A1 [2], COL9A2 [3], and COL9A3 [4,5] can predispose individuals to multiple epiphyseal dysplasia (MED), a clinically highly heterogeneous skeletal disorder with early-onset osteoarthritis (OA) as a very common manifestation. Mice with a completely inactivated COL9A1 gene showed no obvious phenotypic alterations at birth, but thereafter they had a degenerative joint disease resembling human OA at 4 months or older [6]. Transgenic mice lacking type IX collagen and cartilage oligomeric protein are likely to experience cartilage degeneration, which is thought to be mediated by the anchorage of matrilin 3 and proteoglycans. Type IX also controls the diameter of collagen fibrils [7].

Type IX collagen is present in the chondrocytes of growth-plate cartilage, adult articular cartilage, and intervertebral discs. Immunostaining of type IX collagen in normal articular cartilage showed it is concentrated in the pericellular matrix, whereas a weaker intensity appeared in the territorial matrix [8]. In OA tissue higher contents of type IX collagen and its mRNA were found in the pericellular matrix of the weight-bearing areas adjacent to the defects in cartilage [9]. In addition, type IX collagen was reported to be distributed at the edge of fibrillation and fissure in cartilage, suggesting that type IX collagen remodeling is associated with the degeneration of cartilage in OA [8].

All three polypeptide units of type IX collagen, unlike those in fibrillar-forming collagens, contain three triple-helix domains (COL1, COL2, and COL3) interspersed with four non–triple-helix domains (NC1, NC2, NC3, and NC4). Another difference is the longer globular structure of NC4 flanked at the N terminus by α1 chain [10] in type IX. Type IX collagen is located on the surface of collagen fibrils where, in combination with types II and XI collagens, it forms the unique heterofibril network in the matrix of cartilage [11]. NC1 and NC3 contain cysteinyl residues involved in interchain disulfide bond formation. A covalent crosslinking of the NC-1 domain in type IX to type II collagen in cartilage has been detected previously [12]. The N-terminal helical region of the COL2 domain of type IX collagen provides covalent bond sites to the N-telopeptide of α1(II), whereas a covalent bond from the region of COL2 (IX) forms with the C-telopeptide of α1(II) [13]. COL3 and NC4 domains project away from the collagen fibril, and the NC4 domain provides sites for interactions with other matrix components such as glycosaminoglycans and cartilage oligomeric matrix protein (COMP) [14], fibronectin [15], and fibromodulin and osteoadherin [16]. In addition, COL3 and NC4 domains bind to C4, C3 and C9, forming a novel inhibitor of the complement system [17].

Type IX collagen fulfills important functions for cartilage integrity and stability. It also influences the diameter of fibrils through its interaction with COMP at the NC4 domain [7]. In a rabbit model type IX collagen content was significantly reduced in femoral and tibial cartilage at 6 days after antigen ovalbumin-induced rheumatoid arthritis [18]. The immunohistochemical distribution of type IX collagen changed in spontaneously osteoarthritic canine tibial cartilage compared with normal cartilage [8]. These animal models suggest that type IX collagen alterations might be implicated in the pathogenesis of rheumatoid and osteoarthritic arthritis.

BIOMARKERS OF TYPE IX COLLAGEN

Type IX collagen is susceptible to enzymatic degradation. The loss of type IX collagen at the surface of collagen fibrils may make the underlying cartilage vulnerable to proteolytic processing. Release of type IX collagen has been investigated in cartilage in vitro, ex vivo, and in in vivo destruction models. The degradation of type IX collagen was conducted by matrix metalloproteinase (MMP)-3 in vitro and at cleavage sites between 779S and 780L within the NC2 domain [19]. Two major cleavage sites at 258R–259I and 400G–401T, which bound with MMP-13, were identified within the NC4 and the COL3 domains, respectively, in interluekin-1–treated bovine nasal cartilage [20]. Antibodies directed against these neoepitopes might be used to detect degradation caused by proteolytic enzymes in the early stages of OA and rheumatoid arthritis.

Taking all evidence together, type IX collagen is a minor, but crucial, component of collagen fibrils in the cartilage matrix. Turnover of type IX collagen by proteases is an early event in degenerative joint disease. Neoepitopes of type IX collagen degradation might be potential biomarkers for diagnosing and assessing response to treatment in degenerative arthritis.

Type IX Collagen	Description	References
Gene name and number	COL9A1(gene ID: 1297), COL9A2(gene ID: 1298), and COL9A3 (gene ID: 1299)	[1]
Mutations with diseases in man	Predispose for MED phenotypes	[2–5]
Null mutation in mice	Showed no alterations at birth, but associated with degenerative joint disease resembling human osteoarthritis at 4 months or older	[6,7]
Tissue distribution in healthy states	Type IX collagen is present in chondrocytes of growth-plate cartilage, adult articular cartilage, and intervertebral discs Immunostaining of normal articular cartilage showed type IX collagen is concentrated in the pericellular matrix, whereas a weaker intensity appeared in the territorial matrix	[8]
Tissue distribution in pathologically affected states	In the pericellular matrix of the weight-bearing areas adjacent to cartilage defects and at the edge of fibrillation and fissure	[8,9]
Special domains	Three triple-helix domains and four non–triple-helix domains, providing covalent crosslink sites and, especially at the NC4 domain, binding to other matrix components	[12–17]
Special neoepitopes	In vitro, MMP-3 cleaves to a site between 779S and 780L within the NC2 domain Two major cleavage sites at 258R–259I and 400G–401T bound by MMP-13 were identified within the NC4 and COL3 domains, respectively, in IL-1–treated bovine nasal cartilage	[19,20]
Protein structure and function	Each type IX collagen molecule consists of three chains. All three chains contain three triple-helix domains interspersed with four non–triple-helix domains. Type IX collagen in combination with types II and XI collagen form the unique heterofibiril network in the matrix of cartilage	[1,10,11]
Binding proteins	Type II collagen, COMP, fibronectin, fibromodulin, and osteoadherin [16]; also, type IX binds to C4, C3, and C9 to form a novel inhibitor of the complement system	[12,14-17]

Type IX Collagen	Description	References
Known central function	Contributes to the stabilization of the fibrillar collagen network in the cartilage matrix and the anchorage of matrilin 3 and proteoglycans. It controls the diameter of collagen fibrils	[7]
Animals models with protein affected	Ovalbumin-induced rheumatoid arthritis in a rabbit model showed significantly reduced type IX collagen content in femoral and tibial cartilage. Immunohistochemistry showed the distribution of type IX collagen was different in spontaneously osteoarthritic canine tibial cartilage compared with healthy tissue	[8,18]
Biomarkers	A typically used reliable biomarker of type IX collagen has not been reported	

COL, Collagen; *MED*, multiple epiphyseal dysplasia; *MMP*, matrix metalloproteinase; *IL*, interleukin; *COMP*, cartilage oligomeric matrix protein.

REFERENCES

[1] Lozano G, Ninomiya Y, Thompson H, Olsen BR. A distinct class of vertebrate collagen genes encodes chicken type IX collagen polypeptides. Proc Natl Acad Sci USA 1985;82:4050–4.

[2] Czarny-Ratajczak M, Lohiniva J, Rogala P, Kozlowski K, Perala M, Carter L, et al. A mutation in COL9A1 causes multiple epiphyseal dysplasia: further evidence for locus heterogeneity. Am J Hum Genet 2001;69:969–80.

[3] Holden P, Canty EG, Mortier GR, Zabel B, Spranger J, Carr A, et al. Identification of novel pro-alpha2(IX) collagen gene mutations in two families with distinctive oligo-epiphyseal forms of multiple epiphyseal dysplasia. Am J Hum Genet 1999;65:31–8.

[4] Lohiniva J, Paassilta P, Seppanen U, Vierimaa O, Kivirikko S, la-Kokko L. Splicing mutations in the COL3 domain of collagen IX cause multiple epiphyseal dysplasia. Am J Med Genet 2000;90:216–22.

[5] Bonnemann CG, Cox GF, Shapiro F, Wu JJ, Feener CA, Thompson TG, et al. A mutation in the alpha 3 chain of type IX collagen causes autosomal dominant multiple epiphyseal dysplasia with mild myopathy. Proc Natl Acad Sci USA 2000;97:1212–7.

[6] Fassler R, Schnegelsberg PN, Dausman J, Shinya T, Muragaki Y, McCarthy MT, et al. Mice lacking alpha 1 (IX) collagen develop noninflammatory degenerative joint disease. Proc Natl Acad Sci USA 1994;91:5070–4.

[7] Blumbach K, Bastiaansen-Jenniskens YM, DeGroot J, Paulsson M, van Osch GJ, Zaucke F. Combined role of type IX collagen and cartilage oligomeric matrix protein in cartilage matrix assembly: cartilage oligomeric matrix protein counteracts type IX collagen-induced limitation of cartilage collagen fibril growth in mouse chondrocyte cultures. Arthritis Rheum 2009;60:3676–85.

[8] Poole CA, Gilbert RT, Herbage D, Hartmann DJ. Immunolocalization of type IX collagen in normal and spontaneously osteoarthritic canine tibial cartilage and isolated chondrons. Osteoarthritis Cartilage 1997;5:191–204.

[9] Koelling S, Kruegel J, Klinger M, Schultz W, Miosge N. Collagen IX in weight-bearing areas of human articular cartilage in late stages of osteoarthritis. Arch Orthop Trauma Surg 2008;128:1453–9.

Type X Collagen	Description	References
Gene name and number	COL10A1–collagen, type X, alpha 16q21-q22	Gene ID: 1300
Mutations with diseases in humans	More than 30 different mutations have now been characterized. The most common is Schmid metaphyseal chondrodysplasia	[17]
Null mutation in mice	Abnormal trabecular bone, coxa vara, reduction in thickness of the growth plate resting zone and articular cartilage, altered bone content and atypical distribution of matrix components within the cartilage growth plate	[14]
Tissue distribution in healthy states	Hypertrophic zone of the growth plate and basal calcified zone of articular cartilage	[1]
Tissue distribution in pathologically affected states	Increased expression in osteoarthritis when chondrocytes become hypertrophic	[10]
Special domains	Triple helix, NC1 and NC2	[18]
Special neoepitopes	N and C termini	[19,20]
Protein structure and function	A nonfibrillar collagen consisting of three identical α1 chains	[21]
Binding proteins	Anchorin CII	[22]
Known central function	Regulation of endochondral ossification and supporting properties of the growth plate of cartilage and the mineralization process	[3,14]
Animals models with protein affected	None	
Biomarkers	C-COl10	[16]

REFERENCES

[1] Eyre DR. The collagens of articular cartilage. Semin Arthritis Rheum 1991;21:2–11.

[2] Rosati R, Horan GS, Pinero GJ, Garofalo S, Keene DR, Horton WA, et al. Normal long bone growth and development in type X collagen-null mice. Nat Genet 1994;8:129–35.

[3] Shen G. The role of type X collagen in facilitating and regulating endochondral ossification of articular cartilage. Orthod Craniofac Res 2005;8:11–7.

[4] Schmid TM, Mayne R, Jeffrey JJ, Linsenmayer TF. Type X collagen contains two cleavage sites for a vertebrate collagenase. J Biol Chem 1986;261:4184–9.

[5] Gadher SJ, Eyre DR, Duance VC, Wotton SF, Heck LW, Schmid TM, et al. Susceptibility of cartilage collagens type II, IX, X, and XI to human synovial collagenase and neutrophil elastase. Eur J Biochem 1988;175:1–7.

[6] Frischholz S, Beier F, Girkontaite I, Wagner K, Poschl E, Turnay J, et al. Characterization of human type X procollagen and its NC-1 domain expressed as recombinant proteins in HEK293 cells. J Biol Chem 1998;273:4547–55.

[7] Thomas JT, Kwan AP, Grant ME, Boot-Handford RP. Isolation of cDNAs encoding the complete sequence of bovine type X collagen. Evidence for the condensed nature of mammalian type X collagen genes. Biochem J 1991;273(Pt 1):141–8.

[8] Ninomiya Y, Gordon M, van der RM, Schmid T, Linsenmayer T, Olsen BR. The developmentally regulated type X collagen gene contains a long open reading frame without introns. J Biol Chem 1986;261:5041–50.

[9] van der Kraan PM, van den Berg WB. Chondrocyte hypertrophy and osteoarthritis: role in initiation and progression of cartilage degeneration? Osteoarthritis Cartilage 2012;20:223–32.

[10] von der Mark K, Kirsch T, Nerlich A, Kuss A, Weseloh G, Gluckert K, et al. Type X collagen synthesis in human osteoarthritic cartilage. Indication of chondrocyte hypertrophy. Arthritis Rheum 1992;35:806–11.

[11] Lu VP, Iwamoto M, Fanning P, Pacifici M, Olsen BR. Multiple negative elements in a gene that codes for an extracellular matrix protein, collagen X, restrict expression to hypertrophic chondrocytes. J Cell Biol 1993;121:1173–9.

[12] LuValle P, Daniels K, Hay ED, Olsen BR. Type X collagen is transcriptionally activated and specifically localized during sternal cartilage maturation. Matrix 1992;12:404–13.

[13] Brew CJ, Clegg PD, Boot-Handford RP, Andrew JG, Hardingham T. Gene expression in human chondrocytes in late osteoarthritis is changed in both fibrillated and intact cartilage without evidence of generalised chondrocyte hypertrophy. Ann Rheum Dis 2010;69:234–40.

[14] Kwan KM, Pang MK, Zhou S, Cowan SK, Kong RY, Pfordte T, et al. Abnormal compartmentalization of cartilage matrix components in mice lacking collagen X: implications for function. J Cell Biol 1997;136:459–71.

[15] Bohme K, Conscience-Egli M, Tschan T, Winterhalter KH, Bruckner P. Induction of proliferation or hypertrophy of chondrocytes in serum-free culture: the role of insulin-like growth factor-I, insulin, or thyroxine. J Cell Biol 1992;116:1035–42.

[16] He Y, Siebuhr AS, Brandt-Hansen NU, Wang J, Su D, Zheng Q, et al. Type X collagen levels are elevated in serum from human osteoarthritis patients and associated with biomarkers of cartilage degradation and inflammation. BMC Musculoskelet Disord 2014;15:309.

[17] Bateman JF, Wilson R, Freddi S, Lamande SR, Savarirayan R. Mutations of COL10A1 in Schmid metaphyseal chondrodysplasia. Hum Mutat 2005;25:525–34.

[18] Goldring SR, Purdue PE, Crotti TN, Shen Z, Flannery MR, Binder NB, et al. Bone remodelling in inflammatory arthritis. Ann Rheum Dis 2013;72(Suppl. 2):ii52–5.

[19] Ricard-Blum S, Ruggiero F. The collagen superfamily: from the extracellular matrix to the cell membrane. Pathol Biol (Paris) 2005;53:430–42.

[20] Nelson F, Dahlberg L, Laverty S, Reiner A, Pidoux I, Ionescu M, et al. Evidence for altered synthesis of type II collagen in patients with osteoarthritis. J Clin Invest 1998;102:2115–25.

[21] Bruckner P, van der Rest M. Structure and function of cartilage collagens. Microsc Res Tech 1994;28:378–84.

[22] Kirsch T, Pfaffle M. Selective binding of anchorin CII (annexin V) to type II and X collagen and to chondrocalcin (C-propeptide of type II collagen). Implications for anchoring function between matrix vesicles and matrix proteins. FEBS Lett 1992;310:143–7.

Chapter 11

Type XI Collagen

Y.Y. Luo, M.A. Karsdal
Nordic Bioscience, Herlev, Denmark

Chapter Outline

SUMMARY

Type XI collagen is a fibrillary collagen. Type XI collagen is broadly distributed in articular cartilage, testis, trachea, tendons, trabecular bone, skeletal muscle, placenta, lung, and the neoepithelium of the brain. Type XI collagen is able to regulate fibrillogenesis by maintaining the spacing and diameter of type II collagen fibrils, and a nucleator for the fibrillogenesis of collagen types I and II. Mutations in type XI collagen are associated with Stickler syndrome, Marshall syndrome, fibrochondrogenesis, otospondylomegaepiphyseal dysplasia deafness, and Weissenbacher–Zweymüller syndrome. Type XI collagen binds heparin, heparan sulfate, and dermatan sulfate. Currently there are no biomarkers for type XI collagen.

The COL11A1, COL11A2, and COL11A3 genes encode α chains of type XI collagen (COLXI), a member of the fibrillar collagen subgroup [1,2]. COLXI is a minor constituent of type II–containing fibrils and presents an α1(XI)α2(XI)α3(XI) heterotrimer, which is most abundantly found in cartilage. Mutations in the genes are associated with Stickler syndrome, Marshall syndrome, fibrochondrogenesis, otospondylomegaepiphyseal dysplasia (OSMED) deafness, Weissenbacher–Zweymüller syndrome, and nonsyndromic deafness [3–5].

Type XI collagen molecules are primarily crosslinked to each other in cartilage. The crosslinks result in the formation of mature type XI collagen fibers, with the help of collagen types II and IX. Type XI collagen is broadly distributed in articular cartilage, testis, trachea, tendons, trabecular bone, skeletal muscle, placenta, lung, and the neoepithelium of the brain [6,7]. Type XI collagen is also a part of the jelly-like substance that fills the eyeball, the inner ear, and the center portion of the intervertebral discs. However, it is preferentially retained at the chondrocyte surface and involved in the organization of the pericellular matrix via interaction with cartilage proteoglycans [8].

Type XI collagen can form uniformly thin fibrils by acting as the basic structure for the type II collagen fibril network of developing cartilage. It is also the nucleator for the fibrillogenesis of collagen types I and II [9–11]. It is able to regulate fibrillogenesis by maintaining the spacing and diameter of type II collagen fibrils.

It is reported that type XI collagen plays pathogenic roles in musculoskeletal diseases. Deficiencies in this collagen can result in chondrodystrophies in mice and humans. The thrombospondin-like domains of collagen XI extend from the core microfibrils onto the fibril surface [9]. Retained N-propeptide domains are thought to inhibit lateral growth as well as regulate fibril assembly [12].

Type XI collagen is resistant to collagenases, but it is hydrolyzed by gelatinases, giving rise to many peptides. Enzymatic degradation of type XI collagen was believed to play a significant role in the turnover of articular cartilage in both healthy and diseased states [13]. It has been shown that one mutation of type XI collagen leads to increased degradation of type II collagen in articular cartilage [14]. Immunization of rats with homologous type XI collagen was observed to lead to chronic and relapsing arthritis with different genetics and joint pathology than arthritis induced by homologous type II collagen [15]. Type XI collagen–induced arthritis is believed to be associated with the RT1f haplotype, in contrast to type II collagen–induced arthritis, which is associated with the RT1a and RT1u haplotypes [16].

In addition, many functional studies have confirmed that type XI collagen is distributed in a variety of cancerous tissues including breast, lung, colon, and pancreatic tumors. This distribution suggests type XI collagen plays a critical role in cancer proliferation, invasion, and metastasis, as well as in resistance to therapeutics [17]. Levels of ColXIα1 have also been shown to be increased, compared with normal tissues, in several tumor types including colorectal, pancreatic, breast, and non-small-cell lung cancer [18]. It can be used as an accurate marker for differential diagnosis of breast carcinoma invasiveness [19,20]. Collagen XIα1, with its low expression in normal tissue and high expression in cancer, may be an important therapeutic target [21].

Type XI Collagen	Description	References
Gene name and number	COL11A1 (gene ID: 1301) COL11A2 (gene ID: 1302) COL11A3 (ie, COL2A1)	[1,2]
Mutations with diseases in humans	Stickler syndrome Marshall syndrome Fibrochondrogenesis (severe, autosomal-recessive, short-limbed skeletal dysplasia) OSMED Weissenbacher–Zweymüller syndrome Nonsyndromic deafness	[3–5]
Null mutation in mice	Homozygous chondrodysplasia (cho/cho) mouse	[10]

Type XI Collagen	Description	References
Tissue distribution in healthy states	Intervertebral discs, articular cartilage, testis, trachea, tendons, trabecular bone, skeletal muscle, placenta, lung, and neoepithelium of the brain	[6,7]
Tissue distribution in pathologically affected states	Cancerous tissues	[17]
Special domains	Thrombospondin-like, noncollagenous	[9,12,22]
Special neoepitopes	N.A.	
Protein structure and function	Heterotrimeric molecules composed of α1(XI), α2(XI), and α3(XI) subunits Forms uniformly thin fibrils; acts as the basic structure template for the characteristic type II collagen fibril network of developing cartilage; nucleator for the fibrillogenesis of collagen types I and II; maintains the spacing and diameter of type II collagen fibrils; establishes and maintains tissue integrity and cohesion	[9,10,23]
Binding proteins	Heparin, heparan sulfate, and dermatan sulfate	[11]
Known central function	Regulation of fibrillogenesis, such as by controlling the diameter of major collagen fibrils	[11]
Animals models with protein affected	Type XI collagen–induced arthritis in DBA/1 mice; type XI collagen–induced chronic arthritis in the rat	[16,24]
Biomarkers	Splicing patterns of type XI collagen (markers for osteochondrogenic tumors); type XI collagen α1 (accurate marker in the differential diagnosis of breast carcinoma invasiveness in core needle biopsies)	[19,20]

COL, collagen; *N.A.*, not applicable; *OSMED*, otospondylomegaepiphyseal dysplasia deafness.

REFERENCES

[1] Henry I, Bernheim A, Bernard M, van der RM, Kimura T, Jeanpierre C, et al. Mapping of a human fibrillar collagen gene, pro alpha 1 (XI) (COL11A1), to the p21 region of chromosome 1. Genomics July 1988;3(1):87–90.

[2] Kimura T, Cheah KS, Chan SD, Lui VC, Mattei MG, van der Rest M, et al. The human alpha 2(XI) collagen (COL11A2) chain. Molecular cloning of cDNA and genomic DNA reveals characteristics of a fibrillar collagen with differences in genomic organization. J Biol Chem 1989;264(23):13910–6.

[3] Carter EM, Raggio CL. Genetic and orthopedic aspects of collagen disorders. Curr Opin Pediatr February 2009;21(1):46–54.

[4] Tompson SW, Bacino CA, Safina NP, Bober MB, Proud VK, Funari T, et al. Fibrochondrogenesis results from mutations in the COL11A1 type XI collagen gene. Am J Hum Genet November 12, 2010;87(5):708–12.

[5] Sirko-Osadsa DA, Murray MA, Scott JA, Lavery MA, Warman ML, Robin NH. Stickler syndrome without eye involvement is caused by mutations in COL11A2, the gene encoding the alpha2(XI) chain of type XI collagen. J Pediatr February 1998;132(2):368–71.

[6] Mio F, Chiba K, Hirose Y, Kawaguchi Y, Mikami Y, Oya T, et al. A functional polymorphism in COL11A1, which encodes the alpha 1 chain of type XI collagen, is associated with susceptibility to lumbar disc herniation. Am J Hum Genet December 2007;81(6):1271–7.

[7] Yoshioka H, Iyama K, Inoguchi K, Khaleduzzaman M, Ninomiya Y, Ramirez F. Developmental pattern of expression of the mouse alpha 1 (XI) collagen gene (Col11a1). Dev Dyn September 1995;204(1):41–7.

[8] Smith Jr GN, Hasty KA, Brandt KD. Type XI collagen is associated with the chondrocyte surface in suspension culture. Matrix June 1989;9(3):186–92.

[9] Kadler KE, Hill A, Canty-Laird EG. Collagen fibrillogenesis: fibronectin, integrins, and minor collagens as organizers and nucleators. Curr Opin Cell Biol October 2008;20(5):495–501.

[10] Fernandes RJ, Weis M, Scott MA, Seegmiller RE, Eyre DR. Collagen XI chain misassembly in cartilage of the chondrodysplasia (cho) mouse. Matrix Biol October 2007;26(8):597–603.

[11] Vaughan-Thomas A, Young RD, Phillips AC, Duance VC. Characterization of type XI collagen-glycosaminoglycan interactions. J Biol Chem February 16, 2001;276(7):5303–9.

[12] Fallahi A, Kroll B, Warner LR, Oxford RJ, Irwin KM, Mercer LM, et al. Structural model of the amino propeptide of collagen XI alpha1 chain with similarity to the LNS domains. Protein Sci June 2005;14(6):1526–37.

[13] Yu Jr LP, Smith Jr GN, Brandt KD, Capello W. Type XI collagen-degrading activity in human osteoarthritic cartilage. Arthritis Rheum November 1990;33(11):1626–33.

[14] Rodriguez RR, Seegmiller RE, Stark MR, Bridgewater LC. A type XI collagen mutation leads to increased degradation of type II collagen in articular cartilage. Osteoarthritis Cartilage April 2004;12(4):314–20.

[15] Lu S, Carlsen S, Hansson AS, Holmdahl R. Immunization of rats with homologous type XI collagen leads to chronic and relapsing arthritis with different genetics and joint pathology than arthritis induced with homologous type II collagen. J Autoimmun May 2002;18(3):199–211.

[16] Cremer MA, Ye XJ, Terato K, Owens SW, Seyer JM, Kang AH. Type XI collagen-induced arthritis in the Lewis rat. Characterization of cellular and humoral immune responses to native types XI, V, and II collagen and constituent alpha-chains. J Immunol July 15, 1994;153(2):824–32.

[17] Bowen KB, Reimers AP, Luman S, Kronz JD, Fyffe WE, Oxford JT. Immunohistochemical localization of type XI collagen alpha1 and alpha2 chains in human colon tissue. J Histochem Cytochem March 2008;56(3):275–83.

[18] Sok JC, Lee JA, Dasari S, Joyce S, Contrucci SC, Egloff AM, et al. Type XI collagen α1 facilitates head and neck squamous cell cancer growth and invasion. Br J Cancer December 10, 2013;109(12):3049–56.

[19] Freire J, Dominguez-Hormaetxe S, Pereda S, De JA, Vega A, Simon L, et al. Collagen, type XI, alpha 1: an accurate marker for differential diagnosis of breast carcinoma invasiveness in core needle biopsies. Pathol Res Pract December 2014;210(12):879–84.

[20] Matsui Y, Kimura T, Tsumaki N, Nakata K, Yasui N, Araki N, et al. Splicing patterns of type XI collagen transcripts act as molecular markers for osteochondrogenic tumors. Cancer Lett February 27, 1998;124(2):143–8.

[21] Raglow Z, Thomas SM. Tumor matrix protein collagen XIα1 in cancer. Cancer Lett February 28, 2015;357(2):448–53.

[22] Ricard-Blum S. The collagen family. Cold Spring Harb Perspect Biol January 2011;3(1):a004978.

[23] Lui VC, Kong RY, Nicholls J, Cheung AN, Cheah KS. The mRNAs for the three chains of human type XI collagen are widely distributed but not necessarily co-expressed: implications for homotrimeric, heterotrimeric and heterotypic collagen molecules. Biochem J October 15, 1995;311(Pt 2):511–6.

[24] Boissier MC, Chiocchia G, Ronziere MC, Herbage D, Fournier C. Arthritogenicity of minor cartilage collagens (types IX and XI) in mice. Arthritis Rheum January 1990;33(1):1–8.

Chapter 12

Type XII Collagen

T. Manon-Jensen, M.A. Karsdal
Nordic Bioscience, Herlev, Denmark

Chapter Outline

SUMMARY
Type XII collagen is a fibril-associated collagen with interrupted triple helices. Type XII collagen is a component of human skeletal muscle. Type XII collagen has special domains such as the NC3 domain that carry glycosaminoglycan chains, and it interacts with matrix proteins such as decorin, cartilage oligomeric matrix protein, fibromodulin, and tenascin. Col12a null mice display skeletal abnormalities such as shorter and smaller long bones with reduced mechanical strength and altered vertebrae structure. Col12a(–/–) osteoblasts are disorganized and less polarized, with disrupted cell–cell interactions. Recessive and dominant mutations in COL12A1 cause a novel Ehlers–Danlos syndrome/myopathy overlap syndrome in humans and mice. Presently there are no biomarkers of type XII collagen.

Type XII collagen is a fibril-associated collagen with interrupted triple helices. It is a component of human skeletal muscle. One of its functions is to temporarily stabilize type I collagen fibrils at its collagenous domain by preventing the fibrils from permanently crosslinking [1]. By organizing collagen fibrils it maintains bone and muscle integrity [2] and can regulate osteoblast polarity and communication during bone formation [3]. These properties are demonstrated by the mutations in type XII collagen, which cause Ehlers–Danlos syndrome associated with skeletal abnormalities and muscle weakness in mice and humans [2]. Type XII collagen is a homotrimer assembled from three homolog $\alpha 1$ chains encoded by the COLXIIA1 gene. It is composed of the collagenous domain (COL1 and COL2), three N-terminal noncollagenous domains (NC1–NC3) assembled by disulfide bonds, and a short interrupted collagen triple helix near the C terminus [4].

The NC3 domain carrying glycosaminoglycan chains interacts with matrix proteins such as decorin, fibromodulin [5,6], and tenascin [7]. Tenascin plays a functional role in stabilizing bone and muscle structure, and when mutated causes Ehlers–Danlos syndrome. Alternative splicing of the NC3 domain into a short and a long variant generates two isoforms of type XII collagen: type XIIa of 350 kDa and a smaller type XIIb of 220 kDa that lacks glycosaminoglycan chains and resembles type XIV collagen [4]. In humans the most abundant isoform is type XIIa collagen [8–10]. The NC1 domain can be alternatively spliced, generating a different C terminus of type XII collagen, and thus up to four isoforms of type XII collagen could exist [11].

Type XII collagen has functions other than maintaining the integrity of bone and muscle. It may indicate hyaline cartilage development by redifferentiating passaged and primary chondrocytes in the early stages of cartilage tissue formation [12]. In addition, Polacek et al. showed type XII collagen is present in the secretome of human passaged chondrocytes [13], but it was not detected in the secretome of cartilage explant [12]. It may influence cell differentiation as well, because knockout mice show delayed endothelial cell maturation [14].

Another role of type XII collagen may be to stabilize the vascular structure and prevent the formation of atherosclerotic lesions. In vitro, human umbilical vein endothelial cells exposed to fluid shear stress secrete type XII collagen. Supporting data are that when type XII collagen is highly prevalent in the human aortic wall, atherosclerotic lesions are absent, whereas collagen XII is only weakly present in the intima of atherosclerotic plaques [15]. It can only be speculated as to whether this activity against atherosclerotic lesions is due to the von Willebrand factor A–like domain present in type XII collagen. Platelets are major players in the initiation of the atherogenetic process during which the von Willebrand factor is a well-known adhesive mediator between platelets and the vascular wall [16]. Interestingly, patients with hemophilia or von Willebrand disease are believed to be protected from atherosclerosis because of their coagulation defect, although this hypothesis remains controversial [17].

Finally, it has been speculated that type XII collagen is involved in fibrosis and cancer, in particular by aiding cancer cell dissemination. The expression of type XII collagen is considerably higher in the highly malignant breast cancer cell line MCF10CA1 compared with a nonmalignant cell line [18]. In vivo, type XII collagen has been found in the stroma of breast invasive ductal cancer [19] and in the desmoplastic invasive front of colorectal cancer metastasis [20]. It has been found as a marker of myofibroblastic differentiation during colorectal cancer metastasis [20]. Expression of type XII collagen is altered by transforming growth factor-β, a well-known contributor to breast cancer [21,22]. Bleomycin-induced pulmonary fibrosis increases expression of types XII and XIV collagen in mice [23].

Type XII Collagen	Description	References
Gene name and number	COLXIIA1, chromosome 6q12-q13. Transcript variant long, NCBI accession no. NM_004370.5. Transcript variant short, NCBI accession no. NM_080645.2.	[24]
Mutations with diseases in humans	Recessive and dominant mutations in COL12A1 cause a novel EDS/myopathy overlap syndrome in humans and mice. COL12A1 gene polymorphism is associated with anterior cruciate ligament ruptures in women.	[25,26]
Null mutation in mice	COL12A null mice display skeletal abnormalities. They have shorter and smaller long bones with reduced mechanical strength and altered vertebrae structure. COL12A (–/–) osteoblasts have delayed maturation and are disorganized and less polarized, with disrupted cell–cell interactions; decreased connexin43 expression; and impaired gap junction function. Abnormal corneal endothelial maturation in collagens XII and XIV null mice. Recessive and dominant mutations in COL12A1 cause a novel EDS/myopathy overlap syndrome in humans and mice.	[3,14,25]
Tissue distribution in healthy states	Found in association with type I collagen. Both isoforms appear in amnion, chorion, skeletal muscle, small intestine, and in the cell culture of dermal fibroblasts, keratinocytes, and endothelial cells. Only the short isoform is found in the lung, placenta, kidney and a squamous cell carcinoma cell line. Only the long isoform is found in the corneal epithelium, Bowman membrane, and the interfibrillar matrix of the corneal stroma.	[10,27,28]
Tissue distribution in pathologically affected states	In the stroma of IDC breast cancer; in the desmoplastic invasive front of colorectal cancer metastasis.	[19,20]
Special domains	NC1 long and NC3 short splice variants of type XII collagen are overexpressed during corneal scarring.	[29]
Special neoepitopes	N.A.	N/A
Protein structure and function	Type XII is a fibril-associated collagen with interrupted triple helices (FACIT) collagen, which functions to temporarily stabilize type I collagen fibrils as its collagenous domain prevents the fibrils from permanently crosslinking.	[1]

Type XII Collagen	Description	References
Binding proteins	Decorin, fibromodulin, tenascin-X, COMP.	[5–7,30]
Known central function	Protect bone and muscle integrity by organization of collagen fibrils.	[2]
Animals models with protein affected	N.A.	N/A
Biomarkers	N.A.	[20]

COL, Collagen; *COMP*, cartilage oligomeric matrix protein; *EDS*, Ehlers–Danlos syndrome; *FACIT*, fibril-associated collagen with interrupted triple helices; *IDC*, invasive ductal cancer; *N.A.*, not applicable; *NCBI*, National Center for Biotechnology Information.

REFERENCES

[1] Tzortzaki EG, Koutsopoulos AV, Dambaki KI, Lambiri I, Plataki M, Gordon MK, et al. Active remodeling in idiopathic interstitial pneumonias: evaluation of collagen types XII and XIV. J Histochem Cytochem 2006;54:693–700.

[2] Chiquet M, Birk DE, Bonnemann CG, Koch M. Collagen XII: protecting bone and muscle integrity by organizing collagen fibrils. Int J Biochem Cell Biol 2014;53:51–4.

[3] Izu Y, Sun M, Zwolanek D, Veit G, Williams V, Cha B, et al. Type XII collagen regulates osteoblast polarity and communication during bone formation. J Cell Biol 2011;193:1115–30.

[4] Gerecke DR, Olson PF, Koch M, Knoll JH, Taylor R, Hudson DL, et al. Complete primary structure of two splice variants of collagen XII, and assignment of alpha 1(XII) collagen (COL12A1), alpha 1(IX) collagen (COL9A1), and alpha 1(XIX) collagen (COL19A1) to human chromosome 6q12-q13. Genomics 1997;41:236–42.

[5] Font B, Eichenberger D, Goldschmidt D, Boutillon MM, Hulmes DJ. Structural requirements for fibromodulin binding to collagen and the control of type I collagen fibrillogenesis–critical roles for disulphide bonding and the C-terminal region. Eur J Biochem 1998;254:580–7.

[6] Font B, Eichenberger D, Rosenberg LM, van der Rest M. Characterization of the interactions of type XII collagen with two small proteoglycans from fetal bovine tendon, decorin and fibromodulin. Matrix Biol 1996;15:341–8.

[7] Veit G, Hansen U, Keene DR, Bruckner P, Chiquet-Ehrismann R, Chiquet M, et al. Collagen XII interacts with avian tenascin-X through its NC3 domain. J Biol Chem 2006;281:27461–70.

[8] Anderson S, SundarRaj S, Fite D, Wessel H, Sundarraj N. Developmentally regulated appearance of spliced variants of type XII collagen in the cornea. Invest Ophthalmol Vis Sci 2000;41:55–63.

[9] Kabosova A, Azar DT, Bannikov GA, Campbell KP, Durbeej M, Ghohestani RF, et al. Compositional differences between infant and adult human corneal basement membranes. Invest Ophthalmol Vis Sci 2007;48:4989–99.

[10] Wessel H, Anderson S, Fite D, Halvas E, Hempel J, Sundarraj N. Type XII collagen contributes to diversities in human corneal and limbal extracellular matrices. Invest Ophthalmol Vis Sci 1997;38:2408–22.

[11] Kania AM, Reichenberger E, Baur ST, Karimbux NY, Taylor RW, Olsen BR, et al. Structural variation of type XII collagen at its carboxyl-terminal NC1 domain generated by tissue-specific alternative splicing. J Biol Chem 1999;274:22053–9.

[12] Taylor DW, Ahmed N, Parreno J, Lunstrum GP, Gross AE, Diamandis EP, et al. Collagen type XII and versican are present in the early stages of cartilage tissue formation by both redifferentating passaged and primary chondrocytes. Tissue Eng Part A 2015;21:683–93.

[13] Polacek M, Bruun JA, Elvenes J, Figenschau Y, Martinez I. The secretory profiles of cultured human articular chondrocytes and mesenchymal stem cells: implications for autologous cell transplantation strategies. Cell Transplant 2011;20:1381–93.

[14] Hemmavanh C, Koch M, Birk DE, Espana EM. Abnormal corneal endothelial maturation in collagen XII and XIV null mice. Invest Ophthalmol Vis Sci 2013;54:3297–308.

[15] Jin X, Iwasa S, Okada K, Ooi A, Mitsui K, Mitsumata M. Shear stress-induced collagen XII expression is associated with atherogenesis. Biochem Biophys Res Commun 2003;308:152–8.

[16] Luo GP, Ni B, Yang X, Wu YZ. von Willebrand factor: more than a regulator of hemostasis and thrombosis. Acta Haematol 2012;128:158–69.

[17] Bilora F, Zanon E, Petrobelli F, Cavraro M, Prandoni P, Pagnan A, et al. Does hemophilia protect against atherosclerosis? a case-control study. Clin Appl Thromb Hemost 2006;12:193–8.

[18] Yen TY, Haste N, Timpe LC, Litsakos-Cheung C, Yen R, Macher BA. Using a cell line breast cancer progression system to identify biomarker candidates. J Proteomics 2014;96:173–83.

[19] Reddy LA, Mikesh L, Moskulak C, Harvey J, Sherman N, Zigrino P, et al. Host response to human breast Invasive Ductal Carcinoma (IDC) as observed by changes in the stromal proteome. J Proteome Res 2014;13:4739–51.

[20] Karagiannis GS, Petraki C, Prassas I, Saraon P, Musrap N, Dimitromanolakis A, et al. Proteomic signatures of the desmoplastic invasion front reveal collagen type XII as a marker of myofibroblastic differentiation during colorectal cancer metastasis. Oncotarget 2012;3:267–85.

[21] Arai K, Kasashima Y, Kobayashi A, Kuwano A, Yoshihara T. TGF-beta alters collagen XII and XIV mRNA levels in cultured equine tenocytes. Matrix Biol 2002;21:243–50.

[22] Buck MB, Knabbe C. TGF-beta signaling in breast cancer. Ann N Y Acad Sci 2006;1089:119–26.

[23] Tzortzaki EG, Tischfield JA, Sahota A, Siafakas NM, Gordon MK, Gerecke DR. Expression of FACIT collagens XII and XIV during bleomycin-induced pulmonary fibrosis in mice. Anat Rec A Discov Mol Cell Evol Biol 2003;275:1073–80.

[24] Gerecke DR, Meng X, Liu B, Birk DE. Complete primary structure and genomic organization of the mouse Col14a1 gene. Matrix Biol 2004;22:595–601.

[25] Zou Y, Zwolanek D, Izu Y, Gandhy S, Schreiber G, Brockmann K, et al. Recessive and dominant mutations in COL12A1 cause a novel EDS/myopathy overlap syndrome in humans and mice. Hum Mol Genet 2014;23:2339–52.

[26] Posthumus M, September AV, O'Cuinneagain D, van der Merwe W, Schwellnus MP, Collins M. The association between the COL12A1 gene and anterior cruciate ligament ruptures. Br J Sports Med 2010;44:1160–5.

[27] Walchli C, Koch M, Chiquet M, Odermatt BF, Trueb B. Tissue-specific expression of the fibril-associated collagens XII and XIV. J Cell Sci 1994;107(Pt 2):669–81.

[28] Bohme K, Li Y, Oh PS, Olsen BR. Primary structure of the long and short splice variants of mouse collagen XII and their tissue-specific expression during embryonic development. Dev Dyn 1995;204:432–45.

[29] Massoudi D, Malecaze F, Soler V, Butterworth J, Erraud A, Fournie P, et al. NC1 long and NC3 short splice variants of type XII collagen are overexpressed during corneal scarring. Invest Ophthalmol Vis Sci 2012;53:7246–56.

[30] Agarwal P, Zwolanek D, Keene DR, Schulz JN, Blumbach K, Heinegard D, et al. Collagen XII and XIV, new partners of cartilage oligomeric matrix protein in the skin extracellular matrix suprastructure. J Biol Chem 2012;287:22549–59.

Chapter 13

Type XIII Collagen

A.S. Siebuhr, M.A. Karsdal
Nordic Bioscience, Herlev, Denmark

Chapter Outline

SUMMARY

Type XIII collagen is a nonfibrillar, type II trans-membrane collagen. It binds to several proteins such as fibronectin, perlecan, nidogen-2, vitronectin, type IV collagen and $\alpha 1\beta 1$ integrin. Type XIII collagen expression is more pronounced during development and post-natal growth but decreases toward adulthood and is expressed in cells producing connective tissue, with higher expression in certain tumors, in corneal wound healing and renal fibrosis. Loss of type XIII collagen is not lethal, but affects maturation of both the pre-synaptic and post-synaptic specializations of the neuromuscular junctions. Overexpression of the normal type XIII collagen α-chain in cartilage and periosteal osteoblasts results in the development of massive bone overgrowth due to an enhanced osteoblast differentiation capacity. There are currently no biomarkers of type XIII collagen.

Type XIII collagen was first described in 1987 [1]. It is a nonfibrillar, type II trans-membrane collagen. The group of trans-membrane collagens also includes collagen types XXIII and XXV [2,3] and XVII, although XVII is different in structure from the other trans-membrane collagens. Type XIII collagen consists of three identical α1(XIII) chains (*COLXIIIA1*), each of which contains three collagenous sequences, COL1–3, with sizes of 104, 172, and 235 residues, respectively, and noncollagenous domains, NC2–4, with sizes of 34, 22, and 18 residues [4–7]. The N-terminal noncollagenous domain NC1, encompasses a 38-residue cytosolic domain, a 23-residue trans-membrane domain, and the first 60 residues of the noncollagenous extracellular sequences adjacent to the plasma membrane [4]. The α chains fold, unlike other collagens, into homotrimers in an N- to C-terminal direction [8]. The large ectodomain harbors a furin-type endoprotease recognition sequence RRRR↓ from which the ectodomain can be released into the pericellular matrix [5,8].

Biochemistry of Collagens, Laminins and Elastin.

In humans, 10 exons undergo alternative splicing that seems to occur independently of each other. The alternative splicing primarily affects Col1, Col3, NC2, and NC4 [9,10]. The remaining domains, Col2, NC1, and NC3, which are not alternatively spliced, are likely to be of high importance in the function of type XIII collagen due to their stability. However, their exact function is still unknown.

Type XIII collagen expression is more pronounced during development and postnatal growth, but decreases toward adulthood and is expressed in cells producing connective tissue [11–14]. Its expression is induced in certain tumors, in corneal wound healing, and in renal fibrosis [15–17]. Studies of the function of type XIII collagen suggest it is involved in various biological maturation and differentiation processes, some of which are associated with inflammation and vasculogenesis. Type XIII collagen seems to exert its effects through ligand binding, anchoring, and oligomerization. It binds to several proteins such as fibronectin, perlecan, nidogen-2, vitronectin, type IV collagen, and $\alpha 1\beta 1$ integrin. It apparently has a further role in regulating inflammation and immunity and may provide protection against cancer development.

Studies in mice imply that the trans-membrane domain of type XIII collagen is of functional importance, but is not necessary for all functions of this collagen [18]. Studies with Col13a1$^{N/N}$ mice have revealed a role for the N terminus of type XIII collagen in anchoring muscle cells to the basement membrane. These mice have an abnormal plasma membrane–basement membrane interface. Lack of type XIII collagen is not lethal, but maturation of both the presynaptic and postsynaptic specializations of the neuromuscular junctions is compromised [14]. Furthermore, overexpression of the normal collagen type XIII α chain in cartilage and periosteal osteoblasts results in the development of massive bone overgrowth due to an enhanced osteoblast differentiation capacity [13]. In summary, type XIII collagen contributes to differentiation of certain cell types, bone balance, and maturation of neuromuscular junctions.

BIOMARKERS OF TYPE XIII COLLAGEN

Due to the shedding of the ectodomain a neoepitope is released into the extracellular matrix. This neoepitope is not thereafter released to circulation, but upon further degradation of the ectodomain this specific neoepitope could be released to circulation and be a biomarker of type XIII collagen. However, as the level of type XIII collagen in healthy states is low, this biomarker could be useful in cancer, corneal wound healing, and renal fibrosis, as the level of type XIII collagen is increased in these pathological states.

Another potential biomarker could be a measure of autoantibodies to type XIII collagen. Antibodies to type XIII collagen have been found to be increased in sera from Graves ophthalmopathy patients and have been studied as biomarkers in the disease [19]. An enzyme-linked immunosorbent assay (COLXIIIAb) for detection of the autoantigen has been made by Wall and colleagues [19]. Graves patients with active ophthalmopathy had higher levels of COLXIIIAb than Graves patients without ophthalmopathy and controls. This indicates that COLXIIIAb is a biomarker of inflammation of orbital adipose connective tissue [20,21]. In a small study of 10 patients receiving rituximab, no decrease in the biomarker was seen upon treatment [22].

Type XIII collagen	Description	References
Gene name and number	COLXIIIA1, ID:1305 Location: 10q22, exon 41	[23]
Mutations with diseases in humans	N.A.	
Null mutation in mice	Not lethal, but maturation of both the presynaptic and postsynaptic specializations of the neuromuscular junction is compromised	[14]
Tissue distribution in healthy states	Connective tissue–producing cells and in focal adhesions. Often in blood vessels and junctional structures such as neuromuscular structures	[11,14,24]
Tissue distribution in pathologically affected states	During development and postnatal growth and is decreasing toward adulthood. Increased expression in certain tumors, corneal wound healing, and renal fibrosis	[11–14,17]
Special domains	aa 1–44: cytoplasmic aa45–61: transmembrane. Signal anchor for type II collagen aa 62–717: topological domain aa 1–121: NC1 aa 122–216: COl1 aa 217–269: NC2 aa 270–441: Col2 aa 442–463: NC3 aa 464–699: Col3 aa 700–717: NC4	[6,7,10,25,26]
Special neoepitopes	Shedding occurs at aa 61–62. Sequence 62–69: Hfrtaelq	[25]
Protein structure and function	Transmembrane, nonfibrillar collagen Many alternative spliced transcript variants	[11,27,28]
Binding proteins	Fibronectin, perlecan, nidogen-2, vitronectin, collagen IV, $\alpha1\beta1$ integrin	[4,6]
Known central function	Covalent crosslinking collagen	[11]
Animals models with protein affected	Homozygous Col13a1$^{N/N}$	[19]
Biomarkers	Autoantibodies of type XIII collagen are used as biomarkers of Graves ophthalmopathy	

COL, Collagen; *N.A.*, not applicable; *aa*, amino acids.

REFERENCES

[1] Pihlajaniemi T, Myllyla R, Seyer J, Kurkinen M, Prockop DJ. Partial characterization of a low molecular weight human collagen that undergoes alternative splicing. Proc Natl Acad Sci U S A 1987;84:940–4.

[2] Banyard J, Bao L, Zetter BR. Type XXIII collagen, a new transmembrane collagen identified in metastatic tumor cells. J Biol Chem 2003;278:20989–94.

[3] Hashimoto T, Wakabayashi T, Watanabe A, Kowa H, Hosoda R, Nakamura A, et al. CLAC: a novel Alzheimer amyloid plaque component derived from a transmembrane precursor, CLAC-P/collagen type XXV. EMBO J 2002;21:1524–34.

[4] Nykvist P, Tu H, Ivaska J, Kapyla J, Pihlajaniemi T, Heino J. Distinct recognition of collagen subtypes by alpha(1)beta(1) and alpha(2)beta(1) integrins. Alpha(1)beta(1) mediates cell adhesion to type XIII collagen. J Biol Chem 2000;275:8255–61.

[5] Snellman A, Tu H, Vaisanen T, Kvist AP, Huhtala P, Pihlajaniemi T. A short sequence in the N-terminal region is required for the trimerization of type XIII collagen and is conserved in other collagenous transmembrane proteins. EMBO J 2000;19:5051–9.

[6] Tu H, Sasaki T, Snellman A, Gohring W, Pirila P, Timpl R, et al. The type XIII collagen ectodomain is a 150-nm rod and capable of binding to fibronectin, nidogen-2, perlecan, and heparin. J Biol Chem 2002;277:23092–9.

[7] Snellman A, Keranen MR, Hagg PO, Lamberg A, Hiltunen JK, Kivirikko KI, et al. Type XIII collagen forms homotrimers with three triple helical collagenous domains and its association into disulfide-bonded trimers is enhanced by prolyl 4-hydroxylase. J Biol Chem 2000;275:8936–44.

[8] Vaisanen MR, Vaisanen T, Pihlajaniemi T. The shed ectodomain of type XIII collagen affects cell behaviour in a matrix-dependent manner. Biochem J 2004;380:685–93.

[9] Pihlajaniemi T, Tamminen M. The alpha 1 chain of type XIII collagen consists of three collagenous and four noncollagenous domains, and its primary transcript undergoes complex alternative splicing. J Biol Chem 1990;265:16922–8.

[10] Peltonen S, Rehn M, Pihlajaniemi T. Alternative splicing of mouse alphal(XIII) collagen RNAs results in at least 17 different transcripts, predicting alphal(XIII) collagen chains with length varying between 651 and 710 amino acid residues. DNA Cell Biol 1997;16:227–34.

[11] Hagg P, Vaisanen T, Tuomisto A, Rehn M, Tu H, Huhtala P, et al. Type XIII collagen: a novel cell adhesion component present in a range of cell-matrix adhesions and in the intercalated discs between cardiac muscle cells. Matrix Biol 2001;19:727–42.

[12] Sund M, Vaisanen T, Kaukinen S, Ilves M, Tu H, utio-Harmainen H, et al. Distinct expression of type XIII collagen in neuronal structures and other tissues during mouse development. Matrix Biol 2001;20:215–31.

[13] Ylonen R, Kyronlahti T, Sund M, Ilves M, Lehenkari P, Tuukkanen J, et al. Type XIII collagen strongly affects bone formation in transgenic mice. J Bone Miner Res 2005;20:1381–93.

[14] Latvanlehto A, Fox MA, Sormunen R, Tu H, Oikarainen T, Koski A, et al. Muscle-derived collagen XIII regulates maturation of the skeletal neuromuscular junction. J Neurosci 2010;30:12230–41.

[15] Vaisanen T, Vaisanen MR, utio-Harmainen H, Pihlajaniemi T. Type XIII collagen expression is induced during malignant transformation in various epithelial and mesenchymal tumours. J Pathol 2005;207:324–35.

[16] Maatta M, Vaisanen T, Vaisanen MR, Pihlajaniemi T, Tervo T. Altered expression of type XIII collagen in keratoconus and scarred human cornea: Increased expression in scarred cornea is associated with myofibroblast transformation. Cornea 2006;25:448–53.

[17] Dennis J, Meehan DT, Delimont D, Zallocchi M, Perry GA, O'Brien S, et al. Collagen XIII induced in vascular endothelium mediates alpha1beta1 integrin-dependent transmigration of monocytes in renal fibrosis. Am J Pathol 2010;177:2527–40.
[18] Kvist AP, Latvanlehto A, Sund M, Eklund L, Vaisanen T, Hagg P, et al. Lack of cytosolic and transmembrane domains of type XIII collagen results in progressive myopathy. Am J Pathol 2001;159:1581–92.
[19] Yamada M, Li AW, Wall JR. Thyroid-Associated Ophthalmopathy: Clinical Features, Pathogenesis, and Management. Crit Rev Clin Lab Sci 2000;37:523–49.
[20] De BA, Sansone D, Coronella C, Conte M, Iorio S, Perrino S, et al. Serum antibodies to collagen XIII: a further good marker of active Graves' ophthalmopathy. Clin Endocrinol (Oxf) 2005;62:24–9.
[21] Gopinath B, Musselman R, Adams CL, Tani J, Beard N, Wall JR. Study of serum antibodies against three eye muscle antigens and the connective tissue antigen collagen XIII in patients with Graves' disease with and without ophthalmopathy: correlation with clinical features. Thyroid 2006;16:967–74.
[22] Vannucchi G, Campi I, Bonomi M, Covelli D, Dazzi D, Curro N, et al. Rituximab treatment in patients with active Graves' orbitopathy: effects on proinflammatory and humoral immune reactions. Clin Exp Immunol 2010;161:436–43.
[23] Kvist AP, Latvanlehto A, Sund M, Horelli-Kuitunen N, Rehn M, Palotie A, et al. Complete exon-intron organization and chromosomal location of the gene for mouse type XIII collagen (col13a1) and comparison with its human homologue. Matrix Biol 1999;18:261–74.
[24] Vaisanen MR, Vaisanen T, Tu H, Pirila P, Sormunen R, Pihlajaniemi T. The shed ectodomain of type XIII collagen associates with the fibrillar fibronectin matrix and may interfere with its assembly in vitro. Biochem J 2006;393:43–50.
[25] Juvonen M, Sandberg M, Pihlajaniemi T. Patterns of expression of the six alternatively spliced exons affecting the structures of the COL1 and NC2 domains of the alpha 1(XIII) collagen chain in human tissues and cell lines. J Biol Chem 1992;267:24700–7.
[26] Juvonen M, Pihlajaniemi T. Characterization of the spectrum of alternative splicing of alpha 1 (XIII) collagen transcripts in HT-1080 cells and calvarial tissue resulted in identification of two previously unidentified alternatively spliced sequences, one previously unidentified exon, and nine new mRNA variants. J Biol Chem 1992;267:24693–9.
[27] Heikkinen A, Tu H, Pihlajaniemi T. Collagen XIII: a type II transmembrane protein with relevance to musculoskeletal tissues, microvessels and inflammation. Int J Biochem Cell Biol 2012;44:714–7.
[28] Hagg P, Rehn M, Huhtala P, Vaisanen T, Tamminen M, Pihlajaniemi T. Type XIII collagen is identified as a plasma membrane protein. J Biol Chem 1998;273:15590–7.

Chapter 14

Type XIV Collagen

T. Manon-Jensen, M.A. Karsdal
Nordic Bioscience, Herlev, Denmark

Chapter Outline

SUMMARY

Collagen is a fibril-associated collagen with interrupted triple helices found mainly in skin, tendon, cornea, and articular cartilage. It regulates fibrillogenesis by limiting fibril diameter through the prevention of lateral fusion of adjacent fibrils. Type XIV collagen is often present in areas of high mechanical stress, indicating it potentially has a role in maintaining mechanical tissue. Type XIV collagen has three α1 (XIV) chains composed of three noncollagenous (NC1–3) and two collagenous domains (COL1 and COL2) that adhere to fibrillar collagen. Loss-of-function mutation in results in punctuated palmoplantar keratoderma, and mice null for collagen XIV are viable; however, formation of the interstitial collagen network is defective in tendons and skin, leading to reduced biomechanical function. There are no biomarkers available for type XIV collagen.

Type XIV collagen is a fibril-associated collagen with interrupted triple helices (FACIT) found mainly in skin, tendon, cornea, and articular cartilage [1–4]. It regulates fibrillogenesis by limiting fibril diameter through the prevention of lateral fusion of adjacent fibrils [1]. This is unlike another FACIT collagen, type XII collagen, which stabilizes fibrils during development and remodeling [5]. Type XIV collagen has three α1 (XIV) chains composed of three noncollagenous (NC1–3) and two collagenous domains (COL1 and COL2) that adhere to fibrillar collagen [6] (Fig. 4, Introduction). The NC3 domain contains modules structurally similar to von Willebrand factor A–like domains and fibronectin type III repeats that mediate its functional properties [6]. Alternative splicing of the NC1 domain generates two splice variants that are expressed during development and have been suggested to mediate the maturation of fibrils [1].

Type XIV collagen is often present in areas of high mechanical stress, indicating it potentially has a role in maintaining mechanical tissue [2,3]. It has been

Biochemistry of Collagens, Laminins and Elastin.

shown to be important for extracellular matrix assembly and tissue function [7]. Null mice for type XIV collagen have been shown to be viable, but the formation of the interstitial collagen network is defective in tendons and skin, leading to reduced biomechanical function. These mice had defects in fibril growth and fiber assembly during embryonic development [7]. Type XIV, as well as type XII, collagen null mice are also found to have delayed endothelial maturation of the cornea. The structural alterations suggest changes in the endothelium result in increased corneal thickness [8]. It may be speculated that in the fibrotic process type XIV collagen plays a role distinctly different to that of the FACIT type XII collagen. Unlike type XII collagen, type XIV collagen has been found to be present in the later stages of bleomycin-induced mouse lung fibrosis [8,9].

Type XIV Collagen	Collagen Type	References
Gene name and number	COL14A1, chromosome 8q23-q24.1 NCBI accession no.: NM_021110	[10]
Mutations with diseases in humans	Loss-of-function mutation in AAGAB in Chinese families with punctuated palmoplantar keratoderma.	[11]
Null mutation in mice	Abnormal corneal endothelial maturation in collagen types XII and XIV null mice.	[8]
	Mice null for collagen XIV are viable; however, formation of the interstitial collagen network is defective in tendons and skin, leading to reduced biomechanical function.	[7]
Tissue distribution in healthy states	Studies of type XIV collagen in chicken, bovine, and human tissues showed that type XIV collagen is prevalent in skin, tendon, cornea, and articular cartilage	[1–4,12]
Tissue distribution in pathologically affected states	Type XIV collagen is observed in the later stages of fibrosis	[9]
Special domains	The two NC1 splice variants expressed during development may mediate the maturation of fibrils	[1]
Special neoepitopes	Degradation of the COL1 domain of type XIV collagen by 92-kDa gelatinase.	[13]
Protein structure and function	FACIT collagen. It interacts with fibrillar collagens to limit fibril diameter by preventing lateral fusions of adjacent fibrils	[1]
Binding proteins	Possibly decorin and type I collagen	[6,14]
Known central function	Regulates fibrillogenesis by limiting fibril diameter through the prevention of lateral fusion of adjacent fibrils	[1]
Animals models with protein affected	In zebrafish, it is transiently expressed in epithelia and is required for the proper function of certain basement membranes	[15]
Biomarkers	NA	

Col, Collagen; *FACIT*, fibril-associated collagen with interrupted triple helices; *NA*, not applicable; *NCBI*, National Center for Biotechnology Information.

REFERENCES

[1] Young BB, Gordon MK, Birk DE. Expression of type XIV collagen in developing chicken tendons: association with assembly and growth of collagen fibrils. Dev Dyn 2000;217:430–9.
[2] Berthod F, Germain L, Guignard R, Lethias C, Garrone R, Damour O, et al. Differential expression of collagens XII and XIV in human skin and in reconstructed skin. J Invest Dermatol 1997;108:737–42.
[3] Niyibizi C, Visconti CS, Kavalkovich K, Woo SL. Collagens in an adult bovine medial collateral ligament: immunofluorescence localization by confocal microscopy reveals that type XIV collagen predominates at the ligament-bone junction. Matrix Biol 1995;14:743–51.
[4] Walchli C, Koch M, Chiquet M, Odermatt BF, Trueb B. Tissue-specific expression of the fibril-associated collagens XII and XIV. J Cell Sci 1994;107(Pt 2):669–81.
[5] Marchant JK, Zhang G, Birk DE. Association of type XII collagen with regions of increased stability and keratocyte density in the cornea. Exp Eye Res 2002;75:683–94.
[6] Gerecke DR, Meng X, Liu B, Birk DE. Complete primary structure and genomic organization of the mouse Col14a1 gene. Matrix Biol 2004;22:595–601.
[7] Ansorge HL, Meng X, Zhang G, Veit G, Sun M, Klement JF, et al. Type XIV collagen regulates fibrillogenesis: premature collagen fibril growth and tissue dysfunction in null mice. J Biol Chem 2009;284:8427–38.
[8] Hemmavanh C, Koch M, Birk DE, Espana EM. Abnormal corneal endothelial maturation in collagen XII and XIV null mice. Invest Ophthalmol Vis Sci 2013;54:3297–308.
[9] Tzortzaki EG, Tischfield JA, Sahota A, Siafakas NM, Gordon MK, Gerecke DR. Expression of FACIT collagens XII and XIV during bleomycin-induced pulmonary fibrosis in mice. Anat Rec A Discov Mol Cell Evol Biol 2003;275:1073–80.
[10] Imhof M, Trueb B. Comparative cytogenetic mapping of COL14A1, the gene for human and mouse collagen XIV. Cytogenet Cell Genet 1999;84:217–9.
[11] Li M, Yang L, Shi H, Guo B, Dai X, Yao Z, et al. Loss-of-function mutation in AAGAB in Chinese families with punctuate palmoplantar keratoderma. Br J Dermatol 2013;169:168–71.
[12] Lunstrum GP, Morris NP, McDonough AM, Keene DR, Burgeson RE. Identification and partial characterization of two type XII-like collagen molecules. J Cell Biol 1991;113:963–9.
[13] Sires UI, Dublet B, Aubert-Foucher E, van der RM, Welgus HG. Degradation of the COL1 domain of type XIV collagen by 92-kDa gelatinase. J Biol Chem 1995;270:1062–7.
[14] Ehnis T, Dieterich W, Bauer M, Schuppan D. Localization of a cell adhesion site on collagen XIV (undulin). Exp Cell Res 1998;239:477–80.
[15] Bader HL, Lambert E, Guiraud A, Malbouyres M, Driever W, Koch M, et al. Zebrafish collagen XIV is transiently expressed in epithelia and is required for proper function of certain basement membranes. J Biol Chem 2013;288:6777–87.

Chapter 15

Type XV Collagen

A. Arvanitidis, M.A. Karsdal
Nordic Bioscience, Herlev, Denmark

Chapter Outline

SUMMARY

Collagen type XV is from the multiplexin superfamily of collagens, due to the presence of multiple noncollagenous interruptions in their central triple helix. Collagen type XV is predominantly located in the basement membrane zones of microvessels and cardiac or skeletal myocytes. Type XV collagen is produced mainly by fibroblasts, muscle cells, and endothelial cells. To date, there are no known human disease-causing mutations of the COL15A1 gene. However, mice deficient in collagen XV show skeletal myopathy, impaired cardiac function, and defects in the microvasculature of the heart and skin. The fragment of NC1-XV, restin (also named collagen XV endostatin), has ambiguous antiangiogenesis effects, but inhibits endothelial cell migration in vitro. There are presently no biomarkers of type XV collagen.

Collagen type XV is from the multiplexin superfamily of collagens. Collagen type XV is predominantly located in the basement membrane zones of microvessels and cardiac or skeletal myocytes. It is also found in kidney and in interstitial tissues in the pancreas as well as in testis, ovaries, prostate, small intestine, and colon [1]. Type XV collagen is produced mainly by fibroblasts, muscle cells, and endothelial cells [2]. To date, there are no known human disease-causing mutations of the COL15A1 gene. However mice deficient in type XV collagen show skeletal myopathy, impaired cardiac function, and defects in the microvasculature of the heart and skin [3]. Increased deposition of type XV collagen has been observed in fibrotic kidneys and in the sclerotic capillaries of diabetic glomeruli [4]. In addition, loss of type XV collagen from the basement membrane of many tissues has been observed before metastasis of tumors [5].

Type XV collagen shares structural homology with type XVIII collagen. Both are assembled into homotrimers and classified under the multiplexin group due to the presence of multiple noncollagenous interruptions in their central triple helix [6]. Despite both being proteoglycans (PGs), type XVIII collagen is

Biochemistry of Collagens, Laminins and Elastin.

a heparan sulfate PG, whereas type XV collagen has a glycosaminoglycan side chain composition that is variable and consists mainly of chondroitin sulfate. Although type XVIII collagen is abundant in the liver and present only in small amounts in skeletal muscle, collagen XV is absent from the liver and highly expressed in skeletal muscle. Type XVIII collagen endostatin derived from the NC1-XVIII (noncollagenous C terminal) has been shown to possess [7] antimigratory and antitumoral properties. In contrast, the corresponding fragment of NC1-XV, restin (also named type XV collagen endostatin), has ambiguous antiangiogenesis effects, but inhibits endothelial cell migration in vitro [6,8]. Restin is less potent as a tumor suppressor than endostatin, has no posttranslational modifications, and has different binding partners in the extracellular matrix [9] from endostatin. Both collagen types XV and XVIII have been shown to mediate leukocyte influx in renal ischemia and reperfusion [10].

Type XV Collagen	Description	References
Gene name and number	COL15A1	[11]
Mutations with diseases in humans	NA	[1]
Null mutation in mice	Mild skeletal myopathy with increased sensitivity to exercise-induced muscle damage. Cardiovascular problems such as collapsed capillaries and diminished inotropic response	[4]
Tissue distribution in healthy states	Mainly in basement membrane of microvessels or cardiac and skeletal myocytes. Also in kidney, pancreas, intestine, prostate	[6,9]
Tissue distribution in pathologically affected states	Increased deposition in fibrotic kidneys and in the sclerotic capillaries of diabetic glomeruli	[6,8]
Special domains	Restin (also known as XV-endostatin), which is similar to XVIII-endostatin	[6]
Special neoepitopes	C-terminal (NC1-XV) proteolytic production of restin (homologous to endostatin)	[12,13]
Protein structure and function	Proteoglycan. Multiplexin is assembled into homotrimers with multiple noncollagenous interruptions in the central triple helix. Variable glycosaminoglycan side chain composition consisting mainly of chondroitin sulfate. Structural homology to collagen type XVIII	[6]
Binding proteins	Fibulin-2 and nidogen-2, and about 100-fold less to fibulin-1, nidogen-1, laminin-1–nidogen-1 complex, and perlecan	[11]
Known central function	Organization, mechanical stability, and integration of the basement membrane to subjacent connective tissue	[1]
Animals models with protein affected	NA	[4]
Biomarkers	NA	[6,9]

COL, Collagen; *NA*, not applicable.

REFERENCES

[1] Myers JC, Dion AS, Abraham V, Amenta PS. Type XV collagen exhibits a widespread distribution in human tissues but a distinct localization in basement membrane zones. Cell Tissue Res 1996;286:493–505.

[2] Prockop DJ, Kivirikko KI. Collagens: molecular biology, diseases, and potentials for therapy. Annu Rev Biochem 1995;64:403–34.

[3] Rasi K, Piuhola J, Czabanka M, Sormunen R, Ilves M, Leskinen H, et al. Collagen XV is necessary for modeling of the extracellular matrix and its deficiency predisposes to cardiomyopathy. Circ Res 2010;107:1241–52.

[4] Hagg PM, Hagg PO, Peltonen S, Autio-Harmainen H, Pihlajaniemi T. Location of type XV collagen in human tissues and its accumulation in the interstitial matrix of the fibrotic kidney. Am J Pathol 1997;150:2075–86.

[5] Clementz AG, Harris A. Collagen XV: exploring its structure and role within the tumor microenvironment. Mol Cancer Res 2013;11:1481–6.

[6] Iozzo RV. Basement membrane proteoglycans: from cellar to ceiling. Nat Rev Mol Cell Biol 2005;6:646–56.

[7] Amenta PS, Hadad S, Lee MT, Barnard N, Li D, Myers JC. Loss of types XV and XIX collagen precedes basement membrane invasion in ductal carcinoma of the female breast. J Pathol 2003;199:298–308.

[8] Hurskainen M, Ruggiero F, Hagg P, Pihlajaniemi T, Huhtala P. Recombinant human collagen XV regulates cell adhesion and migration. J Biol Chem 2010;285:5258–65.

[9] Ramchandran R, Dhanabal M, Volk R, Waterman MJ, Segal M, Lu H, et al. Antiangiogenic activity of restin, NC10 domain of human collagen XV: comparison to endostatin. Biochem Biophys Res Commun 1999;255:735–9.

[10] Zaferani A, Talsma DT, Yazdani S, Celie JW, Aikio M, Heljasvaara R, et al. Basement membrane zone collagens XV and XVIII/proteoglycans mediate leukocyte influx in renal ischemia/reperfusion. PLoS One 2014;9:e106732.

[11] Eklund L, Piuhola J, Komulainen J, Sormunen R, Ongvarrasopone C, Fassler R, et al. Lack of type XV collagen causes a skeletal myopathy and cardiovascular defects in mice. Proc Natl Acad Sci USA 2001;98:1194–9.

[12] Kobayashi N, Kostka G, Garbe JH, Keene DR, Bachinger HP, Hanisch FG, et al. A comparative analysis of the fibulin protein family. Biochemical characterization, binding interactions, and tissue localization. J Biol Chem 2007;282:11805–16.

[13] Sasaki T, Larsson H, Tisi D, Claesson-Welsh L, Hohenester E, Timpl R. Endostatins derived from collagens XV and XVIII differ in structural and binding properties, tissue distribution and anti-angiogenic activity. J Mol Biol 2000;301:1179–90.

Chapter 16

Type XVI Collagen

J.M.B. Sand, M.A. Karsdal
Nordic Bioscience, Herlev, Denmark

Chapter Outline

SUMMARY

Type XVI collagen is part of the family of fibril-associated collagens with interrupted triple helices that do not form fibrils themselves, but rather associate with fibril-forming collagens as single molecules. Type XVI collagen is synthesized by various cell types including dermal fibroblasts and dendrocytes, keratinocytes, smooth muscle cells, and chondrocytes. It is expressed in several tissues such as skin, cartilage, heart, intestine, arterial walls, and kidney. Type XVI collagen binds to integrins, types II and XI collagen, fibrillin-1, and fibronectin. Its main function is to organize and stabilize the extracellular matrix by stabilizing collagen fibrils and focal adhesions, and anchoring microfibrils to the basement membrane. Furthermore, it mediates intracellular signaling affecting cell adhesion, proliferation, invasiveness, and the formation of focal adhesions. There are currently no mutations reported in man for type XVI collagen, and no mouse null mutations have been reported. There are currently no biomarkers of type XVI collagen.

Type XVI collagen is part of the family of fibril-associated collagens with interrupted triple helices (FACITs) that do not form fibrils themselves, but rather associate with fibril-forming collagens as single molecules [21]. Type XVI collagen was identified in 1992 by two groups in parallel by screening human fibroblast and placenta cDNA libraries. The gene COL16A1 was localized to band p34-35 of chromosome 1 [15,23]. The coding sequence is 1604 amino acids long and contains 10 collagenous domains of 15–422 residues each and 11 non-collagenous (NC) domains, most of which are short (11–39 residues) and cysteine rich [15]. The N-terminal NC11 domain, however, consists of 312 residues, is globular, and contains a proline/arginine-rich protein (PARP) motif also found in several other collagens and in thrombospondin [15,22]. The COL16A1 gene encodes the α1 chain of type XVI collagen, with a molecular mass of approximately 210 kDa, and three α1(XVI) chains associate to form homotrimers [7,11,22]. Cysteine residues primarily located in the NC domains form disulfide bonds that stabilize the trimer [11,22]. Multiple kinks introduced by the NC domains provide a highly flexible protein structure [11].

Type XVI collagen is synthesized by various cell types including dermal fibroblasts and dendrocytes, keratinocytes, smooth muscle cells, and chondrocytes. It is expressed in several tissues such as skin, cartilage, heart, intestine, arterial walls, and kidney [1,2,7,8,10,13,16]. Until now, not much is known about the regulation of gene and protein expression. [13][20]α1(XVI) mRNA levels in human dermal fibroblasts and arterial smooth muscle cells were reduced by basic fibroblast growth factor and elevated by transforming growth factor (TGF)-β2, whereas the effect on protein levels was less pronounced [6,8]. However, stimulation of human coronary smooth muscle cells with TGF-β1 showed no effect on the α1(XVI) mRNA levels [19]. Type XVI collagen is incorporated into distinct tissue-dependent suprastructural structures. In cartilage, type XVI collagen has been identified in the territorial matrix associating with thin weakly banded collagen fibrils containing types II and XI collagen [10]. The FACIT type IX collagen could also be found in these structures, but not in the same locations as type XVI collagen, indicating a mutually exclusive pattern. Thus, the collagen fibrils may be stabilized by either type XVI or IX collagen. Interestingly, type XVI collagen does not seem to be associated with collagen fibrils in skin. Here, type XVI collagen has been localized to the upper papillary dermis in narrow zones near the basement membranes at the dermal–epidermal junction, and to blood vessels where it associates with fibrillin-1 [2,8,10], the major constituent of beaded microfibrils [18]. This association has also been identified in the intestinal wall where type XVI collagen is deposited in the submucosa and colocalizes with both fibrillin-1 and integrin α1 [16]. A remarkable feature of a FACIT collagen is that it not only associates with collagen fibrils but also with non-collagenous proteins such as fibrillin-1. The role of type XVI collagen in this context may be to stabilize the interactions of microfibrils with other matrix components and anchor microfibrils to the basement membrane. Binding of fibronectin to recombinant type XVI collagen has been shown [11], but colocalization in skin was limited to a restricted area near the basement membrane zone, leading the authors to conclude that type XVI was not a component of the fibronectin-containing extracellular matrix (ECM) [8]. Type XVI collagen can associate with cells via integrins α1β1 and α2β1, thus connecting cells with specialized fibrils and contributing to the organization of fibrillar and cellular components of the ECM [5]. Furthermore, it induces integrin-mediated signaling resulting in, for example, increased numbers of focal adhesions, which leads to better cell spreading [16]. Type XVI collagen has been identified as a component of the dorsal root ganglia ECM in developing mice, but it was only found in low levels in adult mice [9,13]. However, its expression in adult mice increases around neuronal cell bodies in response to nerve injury, suggesting a role in the regeneration process [9].

Although type XVI collagen is a minor type of collagen it has been implicated in several diseases of the ECM. Dermal fibroblasts of patients with localized and systemic scleroderma showed type XVI collagen mRNA levels that were elevated 2.3- and 3.6-fold, respectively, compared with healthy controls [2]. Furthermore, deposition of type XVI collagen extended from the upper dermis to include the lower dermal matrix in the skin of patients with systemic scleroderma, whereas it was only found in the superficial dermis of control skin [2]. Epidermolysis bullosa cell lines also showed significantly upregulated mRNA levels, further establishing the role of type XVI collagen in skin diseases [12]. Type XVI collagen expression is affected in patients with Crohn's disease, which is characterized by chronic inflammation of the gastrointestinal tract leading to excessive tissue repair and fibrosis. Intestinal subepithelial myofibroblasts (ISEMFs) from the inflamed bowel wall showed elevated type XVI gene and protein expression compared with ISEMF from noninflamed tissue [16]. The expression of integrin $\alpha1\beta1$ on the surface of ISEMFs originating from inflamed bowel wall was also elevated, resulting in increased numbers of focal adhesion contacts with type XVI collagen [16]. Thus, type XVI collagen may contribute to sustaining the fibrotic response by retaining the collagen-producing ISEMFs in the inflamed tissue. Cancer cells are known to alter the surrounding ECM structure, and evidence suggests that this includes type XVI collagen. Overexpression of type XVI collagen in the superficial epithelial layers of dysplastic areas of the mucosa, but reduced expression in the basement membrane, was seen in patients with oral squamous cell carcinoma (OSCC), whereas expression was restricted to the basement membrane at the dermal–epidermal junction in normal mucosa [17]. Overexpression of type XVI collagen in an OSCC cell line has been shown to induce cell proliferation and elevated expression of matrix metalloproteinase (MMP)-9, resulting from the type XVI collagen–mediated activation of integrin $\beta1$ and kindlin-1 and the subsequent intracellular signaling via integrin-linked kinase activation [4,17]. Furthermore, a glioblastoma cell line as well as human glioblastoma tissue showed overexpression of type XVI collagen mRNA and protein compared with normal cells and cortex tissue [20]. A type XVI collagen substrate induced increased adhesion of glioblastoma cells, whereas cell migration was unchanged [20]. Knockdown of the endogenous protein expression in a glioblastoma cell line reduced glioblastoma cell invasiveness and number of focal adhesions as well as integrin activation, but it did not affect adhesion capacity or proliferation [3]. However, another study using the same cell line showed reduced cell adhesion in response to type XVI collagen knockdown [20]. Overexpression of type XVI collagen has also been observed in two mouse models of hepatocellular carcinoma [14].

BIOMARKERS OF TYPE XVI COLLAGEN

As type XVI collagen has been implicated in several diseases the possibility of using it as a drug target or biomarker exists. Several cleavage sites and resulting fragments of type XVI collagen have been described, suggesting extensive processing of type XVI collagen after secretion [7,11]. Future studies may reveal whether any of these fragments are as biologically active as the known matrikines released from several other collagens. The induction of MMP-9 expression described in OSCC [4] may result in further cleavage of type XVI collagen, possibly releasing disease-specific protein fragments into the circulation. The role of type XVI collagen in ECM organization and stability as well as cell behavior will surely encourage future investigations into its role in diseases.

Type XVI Collagen	Description	References
Gene name and number	COL16A1 (1p35-p34)	NCBI gene ID: 1307 [15,23]
Mutations with diseases in humans	None identified	
Null mutation in mice	N.A.	
Tissue distribution in healthy states	Skin, cartilage, heart, intestine, arterial walls, and kidney; fetal brain, and skeletal muscle.	[1,2,7,8,10,13,16]
Tissue distribution in pathologically affected states	• Overexpression in localized and systemic scleroderma, extending from the papillary dermis to the lower dermis. • Overexpression in the intestinal wall in Crohn's disease. • Overexpression in cancers including glioblastoma, OSCC (but reduced in basement membrane), and HCC. • Overexpression in response to nerve injury.	[2,9,14,16,17,20]
Special domains	COL1–10, NC1–11, PARP/TSPN	[15,22]
Special neoepitopes	• N-terminal processing results in 182- and 78-kDa fragments, cleaved at R256-D257 and E940-L941, respectively. • C-terminal and subsequent N-terminal processing results in 150-, 110-, 50-, and 35-kDa fragments.	[7,11]
Protein structure and function	• FACIT type collagen with kinks resulting in flexibility. • Associated with thin collagen fibrils in cartilage territorial matrix. • Associated with microfibrils near basement membranes in the papillary dermis and intestinal wall submucosa.	[8,10,11,16]

Type XVI Collagen	Description	References
Binding proteins	Integrins $\alpha1\beta1$ and $\alpha2\beta1$, microfibrils, thin weakly banded collagen fibrils (containing types II and XI collagen), fibrillin-1; possibly fibronectin.	[5,10,11]
Known central function	Organizing and stabilizing ECM by stabilizing collagen fibrils and focal adhesions, and anchoring microfibrils to the basement membrane; mediating intracellular signaling affecting cell adhesion, proliferation, invasiveness, and the formation of focal adhesions.	[3,10,16,17,20]
Animals models with protein affected	HCC mouse models: PDGFC transgenic and Pten null mice.	[14]
Biomarkers	N.A.	

COL, collagen; *ECM*, extracellular matrix; *FACIT*, fibril-associated collagens with interrupted triple helices; *HCC*, hepatocellular carcinoma; *NCBI*, National Center for Biotechnology Information; *OSCC*, oral squamous cell carcinoma; *PARP*, proline-arginine-rich protein; *TSPN*, thrombospondin N- terminal–like domain.

REFERENCES

[1] Akagi A, Tajima S, Ishibashi A, Matsubara Y, Takehana M, Kobayashi S, et al. Type XVI collagen is expressed in factor XIIIa+monocyte-derived dermal dendrocytes and constitutes a potential substrate for factor XIIIa. J Invest Dermatol 2002;118:267–74.

[2] Akagi A, Tajima S, Ishibashi A, Yamaguchi N, Nagai Y. Expression of type XVI collagen in human skin fibroblasts: enhanced expression in fibrotic skin diseases. J Invest Dermatol 1999;113:246–50.

[3] Bauer R, Ratzinger S, Wales L, Bosserhoff A, Senner V, Grifka J, et al. Inhibition of collagen XVI expression reduces glioma cell invasiveness. Cell Physiol Biochem 2011;27:217–26.

[4] Bedal KB, Grassel S, Oefner PJ, Reinders J, Reichert TE, Bauer R. Collagen XVI induces expression of MMP9 via modulation of AP-1 transcription factors and facilitates invasion of oral squamous cell carcinoma. PLoS One 2014;9:e86777.

[5] Eble JA, Kassner A, Niland S, Morgelin M, Grifka J, Grassel S. Collagen XVI harbors an integrin alpha1 beta1 recognition site in its C-terminal domains. J Biol Chem 2006;281:25745–56.

[6] Grassel S, Tan EM, Timpl R, Chu ML. Collagen type XVI expression is modulated by basic fibroblast growth factor and transforming growth factor-beta. FEBS Lett 1998;436:197–201.

[7] Grassel S, Timpl R, Tan EM, Chu ML. Biosynthesis and processing of type XVI collagen in human fibroblasts and smooth muscle cells. Eur J Biochem 1996;242:576–84.

[8] Grassel S, Unsold C, Schacke H, Bruckner-Tuderman L, Bruckner P. Collagen XVI is expressed by human dermal fibroblasts and keratinocytes and is associated with the microfibrillar apparatus in the upper papillary dermis. Matrix Biol 1999;18:309–17.

[9] Hubert T, Grimal S, Ratzinger S, Mechaly I, Grassel S, Fichard-Carroll A. Collagen XVI is a neural component of the developing and regenerating dorsal root ganglia extracellular matrix. Matrix Biol 2007;26:206–10.

[10] Kassner A, Hansen U, Miosge N, Reinhardt DP, Aigner T, Bruckner-Tuderman L, et al. Discrete integration of collagen XVI into tissue-specific collagen fibrils or beaded microfibrils. Matrix Biol 2003;22:131–43.

[11] Kassner A, Tiedemann K, Notbohm H, Ludwig T, Morgelin M, Reinhardt DP, et al. Molecular structure and interaction of recombinant human type XVI collagen. J Mol Biol 2004;339:835–53.

[12] Knaup J, Verwanger T, Gruber C, Ziegler V, Bauer J, Krammer B. Epidermolysis bullosa – a group of skin diseases with different causes but commonalities in gene expression. Exp Dermatol 2012;21:526–30.

[13] Lai CH, Chu ML. Tissue distribution and developmental expression of type XVI collagen in the mouse. Tissue Cell 1996;28:155–64.

[14] Lai KK, Shang S, Lohia N, Booth GC, Masse DJ, Fausto N, et al. Extracellular matrix dynamics in hepatocarcinogenesis: a comparative proteomics study of PDGFC transgenic and Pten null mouse models. PLoS Genet 2011;7:e1002147.

[15] Pan TC, Zhang RZ, Mattei MG, Timpl R, Chu ML. Cloning and chromosomal location of human alpha 1(XVI) collagen. Proc Natl Acad Sci USA 1992;89:6565–9.

[16] Ratzinger S, Eble JA, Pasoldt A, Opolka A, Rogler G, Grifka J, et al. Collagen XVI induces formation of focal contacts on intestinal myofibroblasts isolated from the normal and inflamed intestinal tract. Matrix Biol 2010;29:177–93.

[17] Ratzinger S, Grassel S, Dowejko A, Reichert TE, Bauer RJ. Induction of type XVI collagen expression facilitates proliferation of oral cancer cells. Matrix Biol 2011;30:118–25.

[18] Sakai LY, Keene DR, Engvall E. Fibrillin, a new 350-kD glycoprotein, is a component of extracellular microfibrils. J Cell Biol 1986;103:2499–509.

[19] Schmidt A, Lorkowski S, Seidler D, Breithardt G, Buddecke E. TGF-beta1 generates a specific multicomponent extracellular matrix in human coronary SMC. Eur J Clin Invest 2006;36:473–82.

[20] Senner V, Ratzinger S, Mertsch S, Grassel S, Paulus W. Collagen XVI expression is upregulated in glioblastomas and promotes tumor cell adhesion. FEBS Lett 2008;582:3293–300.

[21] Shaw LM, Olsen BR. FACIT collagens: diverse molecular bridges in extracellular matrices. Trends Biochem Sci 1991;16:191–4.

[22] Tillet E, Mann K, Nischt R, Pan TC, Chu ML, Timpl R. Recombinant analysis of human alpha 1 (XVI) collagen. Evidence for processing of the N-terminal globular domain. Eur J Biochem 1995;228:160–8.

[23] Yamaguchi N, Kimura S, McBride OW, Hori H, Yamada Y, Kanamori T, et al. Molecular cloning and partial characterization of a novel collagen chain, alpha 1(XVI), consisting of repetitive collagenous domains and cysteine-containing non-collagenous segments. J Biochem 1992;112:856–63.

Chapter 17

Type XVII Collagen

S. Sun, M.A. Karsdal
Nordic Bioscience, Herlev, Denmark

Chapter Outline

SUMMARY

Type XVII collagen, also known as 180-kDa bullous pemphigoid antigen, is a type II transmembrane collagen. Type XVII collagen is expressed in epithelial hemidesmosomes of skin, mucous membrane, and eye and has a series of binding partners. The intracellular domain binds to β4 integrins, bullous pemphigoid antigen, and plectin, whereas the extracellular domain binds to laminin. The putative function of type XVII collagen is to stabilize adhesion of epithelial cells to the surrounding extracellular matrix. Mutations in type XVII collagen can cause junctional epidermolysis bullosa, with abnormal tooth formation and epithelial recurrent erosion dystrophy disease. Mouse models display similar skin symptoms to those of humans, such as blisters and erosions. Enamel hypoplasia is also found in this model, suggesting type XVII collagen also has important roles in teeth formation. There are currently no biomarkers for type XVII collagen.

Type XVII collagen, also known as 180-kDa bullous pemphigoid antigen (BP180), is a type II transmembrane collagen [1]. It is expressed in epithelial hemidesmosomes of skin, mucous membrane, and eye [2]. Type XVII collagen is composed of three identical 180-kDa α1 chains [3]. The protein structure includes a globular cytoplasmic domain, a transmembrane domain, and an extracellular domain with 15 collagenous domains separated by 16 noncollagenous domains [4]. The intracellular domain, transmembrane domain, and extracellular domain include 466, 23, 1008 amino acids, respectively [1]. Col15 domain, the largest collagenous area, is a cell adhesion domain. The noncollagenous NC16a domain is important for the triple-helix folding [5,6].

Type XVII collagen has a series of binding partners. The intracellular domain binds to β4 integrins, 230-kDa bullous pemphigoid antigen (BP230), and plectin, whereas the extracellular domain binds to laminin 332 [7–10]. Therefore, it is suggested that the function of type XVII collagen is to stabilize adhesion of epithelial cells to the surrounding extracellular matrix (ECM) [1].

Biochemistry of Collagens, Laminins and Elastin.

The extracellular domain of type XVII collagen can be shed from the cell surface by A Disintegrin and Metalloproteinases (ie, ADAMs) and tumor necrosis factor-α–converting enzyme [3,11,12], and release a soluble domain that is usually referred to as 120-kDa linear immunoglobulin (Ig)A dermatosis antigen (LAD-1) [13,14]. However, the precise cleavage site on the collagen is unclear since studies indicate a different N terminus for the shedding: Ala^{528} from HaCaT cells; Leu^{524} from DJM-1 cells; and Leu^{524}, Gln^{525}, and Gly^{526} from an in vitro primary normal human keratinocyte experiment [15,16]. A study showed the main physiological cleavage site in human skin could be between Leu^{524} and Gln^{525} [17]. Besides the 120-kDa fragment, a 97-kDa linear IgA dermatosis antigen (97-LAD) from the ectodomain of type XVII collagen is also found in the epidermis, with the N terminus starting from amino acid 531 [18,19]. The effect of shedding type XVII collagen is still unclear, but potentially, it may release the cell from the surrounding matrix and allow it to perform other functions [1].

Autoantibodies against type XVII collagen are found in autoimmune bullous skin disease. Studies have shown that autoantibodies in serum from some autoimmune bullous pemphigoid patients predominantly react with the NC16a or COL15 domain of type XVII collagen [5,20–22]. Linear IgA disease is another pemphigoid-like illness and characterized as a subepidermal blistering disorder. In these patients, IgA autoantibodies preferentially target LAD-1 and 97-LAD instead of the full-length type XVII collagen [22].

Mutations in type XVII collagen can cause junctional epidermolysis bullosa (JEB). Abnormal tooth formation is also found in some JEB patients [2,23–29]. In JEB, mutations in type XVII collagen result in decreased levels or absence of this protein [30]. In epithelial recurrent erosion dystrophy disease, a novel mutation in type XVII collagen has been identified [31]. Type XVII collagen functions have been further investigated in the COL17 knockout mouse model. The knockout mice show similar skin symptoms to those of humans, such as blisters and erosions [32]. Enamel hypoplasia is also found in this model, suggesting type XVII collagen also has important roles in teeth formation [33].

Type XVII collagen	Description	References
Gene name and number	COL17A1, location 10q24.3 (gene ID: 1308)	NCBI
Mutations with diseases in humans	Junctional epidermolysis bullosa; abnormal tooth formation; epithelial recurrent erosion dystrophy	[2,23–29,31]
Null mutation in mice	Blisters and erosions on the skin; enamel hypoplasia	[32,33]
Tissue distribution in healthy states	Expressed in epithelial hemidesmosomes of skin, mucous membrane, and eye	[2]
Tissue distribution in pathologically affected states	In junctional epidermolysis bullosa, mutations in type XVII collagen also result in decreased levels or absence of this collagen.	[30]
Special domains	LAD-1, 97-LAD	[13,14,18,19]

Type XVII collagen	Description	References
Special neoepitopes	The cleavage sites for LAD-1 and 97-LAD fragments	[15,18]
Protein structure and function	Composed of three identical α1 chains. The collagen includes a globular intracellular domain, a transmembrane domain, and an extracellular domain that is composed of 15 collagenous domains and 16 noncollagenous domains. Col15 domain, the largest collagenous region, is a cell adhesion domain. The noncollagenous NC16a domain is important for the triple-helix folding.	[3–6]
Binding proteins	The intracellular domain binds to β4 integrins, 230-kDa bullous pemphigoid antigen (BP230) and plectin, whereas the extracellular domain binds to laminin 332.	[7–10]
Known central function	Stable adhesion of epithelial cells to the surrounding ECM	[1]
Animal models with protein affected	Col17 knockout mouse	[32,33]
Biomarkers	N.A.	

COL, Collagen; *NCBI*, National Center for Biotechnology Information; *LAD*, linear immunoglobulin A dermatosis antigen; *BP*, bullous pemphigoid antigen; *ECM*, extracellular matrix; *N.A.*, not applicable.

REFERENCES

[1] Franzke CW, Bruckner P, Bruckner-Tuderman L. Collagenous transmembrane proteins: recent insights into biology and pathology. J Biol Chem 2005;280:4005–8.

[2] McGrath JA, Gatalica B, Christiano AM, Li K, Owaribe K, McMillan JR, et al. Mutations in the 180-kD bullous pemphigoid antigen (BPAG2), a hemidesmosomal transmembrane collagen (COL17A1), in generalized atrophic benign epidermolysis bullosa. Nat Genet 1995;11:83–6.

[3] Franzke CW, Tasanen K, Borradori L, Huotari V, Bruckner-Tuderman L. Shedding of collagen XVII/BP180: structural motifs influence cleavage from cell surface. J Biol Chem 2004;279:24521–9.

[4] Franzke CW, Tasanen K, Schumann H, Bruckner-Tuderman L. Collagenous transmembrane proteins: collagen XVII as a prototype. Matrix Biol 2003;22:299–309.

[5] Tasanen K, Eble JA, Aumailley M, Schumann H, Baetge J, Tu H, et al. Collagen XVII is destabilized by a glycine substitution mutation in the cell adhesion domain Col15. J Biol Chem 2000;275:3093–9.

[6] Areida SK, Reinhardt DP, Muller PK, Fietzek PP, Kowitz J, Marinkovich MP, et al. Properties of the collagen type XVII ectodomain. Evidence for n- to c-terminal triple helix folding. J Biol Chem 2001;276:1594–601.

[7] Hopkinson SB, Baker SE, Jones JC. Molecular genetic studies of a human epidermal autoantigen (the 180-kD bullous pemphigoid antigen/BP180): identification of functionally important sequences within the BP180 molecule and evidence for an interaction between BP180 and alpha 6 integrin. J Cell Biol 1995;130:117–25.

[8] Aho S, Uitto J. Direct interaction between the intracellular domains of bullous pemphigoid antigen 2 (BP180) and beta 4 integrin, hemidesmosomal components of basal keratinocytes. Biochem Biophys Res Commun 1998;243:694–9.
[9] Koster J, Geerts D, Favre B, Borradori L, Sonnenberg A. Analysis of the interactions between BP180, BP230, plectin and the integrin alpha6beta4 important for hemidesmosome assembly. J Cell Sci 2003;116:387–99.
[10] Van den BF, Eliason SL, Giudice GJ. Type XVII collagen (BP180) can function as a cell-matrix adhesion molecule via binding to laminin 332. Matrix Biol 2011;30:100–8.
[11] Franzke CW, Bruckner-Tuderman L, Blobel CP. Shedding of collagen XVII/BP180 in skin depends on both ADAM10 and ADAM9. J Biol Chem 2009;284:23386–96.
[12] Franzke CW, Tasanen K, Schacke H, Zhou Z, Tryggvason K, Mauch C, et al. Transmembrane collagen XVII, an epithelial adhesion protein, is shed from the cell surface by ADAMs. EMBO J 2002;21:5026–35.
[13] Pas HH, Kloosterhuis GJ, Nijenhuis M, de Jong MC, van der Meer JB, Jonkman MF. Type XVII collagen (BP180) and LAD-1 are present as separate trimeric complexes. J Invest Dermatol 1999;112:58–61.
[14] Marinkovich MP, Taylor TB, Keene DR, Burgeson RE, Zone JJ. LAD-1, the linear IgA bullous dermatosis autoantigen, is a novel 120-kDa anchoring filament protein synthesized by epidermal cells. J Invest Dermatol 1996;106:734–8.
[15] Hirako Y, Nishizawa Y, Sitaru C, Opitz A, Marcus K, Meyer HE, et al. The 97-kDa (LABD97) and 120-kDa (LAD-1) fragments of bullous pemphigoid antigen 180/type XVII collagen have different N-termini. J Invest Dermatol 2003;121:1554–6.
[16] Nishie W, Lamer S, Schlosser A, Licarete E, Franzke CW, Hofmann SC, et al. Ectodomain shedding generates Neoepitopes on collagen XVII, the major autoantigen for bullous pemphigoid. J Immunol 2010;185:4938–47.
[17] Nishie W, Natsuga K, Iwata H, Izumi K, Ujiie H, Toyonaga E, et al. Context-dependent regulation of collagen XVII ectodomain shedding in skin. Am J Pathol 2015;185:1361–71.
[18] Zone JJ, Taylor TB, Meyer LJ, Petersen MJ. The 97 kDa linear IgA bullous disease antigen is identical to a portion of the extracellular domain of the 180 kDa bullous pemphigoid antigen, BPAg2. J Invest Dermatol 1998;110:207–10.
[19] Ishiko A, Shimizu H, Masunaga T, Hashimoto T, Dmochowski M, Wojnarowska F, et al. 97-kDa linear IgA bullous dermatosis (LAD) antigen localizes to the lamina lucida of the epidermal basement membrane. J Invest Dermatol 1996;106:739–43.
[20] Zillikens D, Rose PA, Balding SD, Liu Z, Olague-Marchan M, Diaz LA, et al. Tight clustering of extracellular BP180 epitopes recognized by bullous pemphigoid autoantibodies. J Invest Dermatol 1997;109:573–9.
[21] Matsumura K, Amagai M, Nishikawa T, Hashimoto T. The majority of bullous pemphigoid and herpes gestationis serum samples react with the NC16a domain of the 180-kDa bullous pemphigoid antigen. Arch Dermatol Res 1996;288:507–9.
[22] Schumann H, Baetge J, Tasanen K, Wojnarowska F, Schacke H, Zillikens D, et al. The shed ectodomain of collagen XVII/BP180 is targeted by autoantibodies in different blistering skin diseases. Am J Pathol 2000;156:685–95.
[23] Aumailley M, Has C, Tunggal L, Bruckner-Tuderman L. Molecular basis of inherited skin-blistering disorders, and therapeutic implications. Expert Rev Mol Med 2006;8:1–21.
[24] Schumann H, Hammami-Hauasli N, Pulkkinen L, Mauviel A, Kuster W, Luthi U, et al. Three novel homozygous point mutations and a new polymorphism in the COL17A1 gene: relation to biological and clinical phenotypes of junctional epidermolysis bullosa. Am J Hum Genet 1997;60:1344–53.

[25] Chavanas S, Gache Y, Tadini G, Pulkkinen L, Uitto J, Ortonne JP, et al. A homozygous in-frame deletion in the collagenous domain of bullous pemphigoid antigen BP180 (type XVII collagen) causes generalized atrophic benign epidermolysis bullosa. J Invest Dermatol 1997;109:74–8.

[26] Darling TN, McGrath JA, Yee C, Gatalica B, Hametner R, Bauer JW, et al. Premature termination codons are present on both alleles of the bullous pemphigoid antigen 2/type XVII collagen gene in five Austrian families with generalized atrophic benign epidermolysis bullosa. J Invest Dermatol 1997;108:463–8.

[27] McGrath JA, Darling T, Gatalica B, Pohla-Gubo G, Hintner H, Christiano AM, et al. A homozygous deletion mutation in the gene encoding the 180-kDa bullous pemphigoid antigen (BPAG2) in a family with generalized atrophic benign epidermolysis bullosa. J Invest Dermatol 1996;106:771–4.

[28] Nakamura H, Sawamura D, Goto M, Nakamura H, Kida M, Ariga T, et al. Analysis of the COL17A1 in non-Herlitz junctional epidermolysis bullosa and amelogenesis imperfecta. Int J Mol Med 2006;18:333–7.

[29] Murrell DF, Pasmooij AM, Pas HH, Marr P, Klingberg S, Pfendner E, et al. Retrospective diagnosis of fatal BP180-deficient non-Herlitz junctional epidermolysis bullosa suggested by immunofluorescence (IF) antigen-mapping of parental carriers bearing enamel defects. J Invest Dermatol 2007;127:1772–5.

[30] Darling TN, Bauer JW, Hintner H, Yancey KB. Generalized atrophic benign epidermolysis bullosa. Adv Dermatol 1997;13:87–119.

[31] Jonsson F, Bystrom B, Davidson AE, Backman LJ, Kellgren TG, Tuft SJ, et al. Mutations in collagen, type XVII, alpha 1 (COL17A1) cause epithelial recurrent erosion dystrophy (ERED). Hum Mutat 2015;36:463–73.

[32] Nishie W, Sawamura D, Goto M, Ito K, Shibaki A, McMillan JR, et al. Humanization of autoantigen. Nat Med 2007;13:378–83.

[33] Asaka T, Akiyama M, Domon T, Nishie W, Natsuga K, Fujita Y, et al. Type XVII collagen is a key player in tooth enamel formation. Am J Pathol 2009;174:91–100.

Chapter 18

Type XVIII Collagen

C.L. Bager, M.A. Karsdal
Nordic Bioscience, Herlev, Denmark

Chapter Outline

SUMMARY

Type XVIII collagen possesses features of collagens and proteoglycans and is localized in various basement membrane zones. Endostatin, the carboxyl-terminal fragment of type XVIII collagen, has been found to inhibit angiogenesis and tumor growth. There are three isoforms of type XVIII collagen: short, middle, and long. The short isoform is present in vascular and epithelial basement membrane structures, and the long isoform is highly expressed in the liver. Nonsense and missense mutations of type XVIII collagen lead to the autosomal recessive disorder Knobloch syndrome-1, which is characterized by eye abnormalities. COL18A1−/− knockout mice likewise suffer from eye abnormalities and manifest broadened basement membrane structures. This suggests that type XVIII collagen not only is involved in eye development but also has an important role in maintaining basement membrane integrity. Type XVIII collagen is the prime example of a collagen that holds structural properties in the intact form and signaling potential in the degraded form.

Abbreviations

Col Collagenous domain
HSGAG Heparan sulphate glycosaminoglycan
NC Noncollagenous region
TSP-1 Thrombospondin 1

Type XVIII collagen possesses features of collagens and proteoglycans and is, together with collagen XV, part of the multiplexin family [1]. Multiplexins differ from other known collagens by having triple helical domains flanked by unique nontriple helical regions, which possibly allows flexibility in the collagen molecule [2].

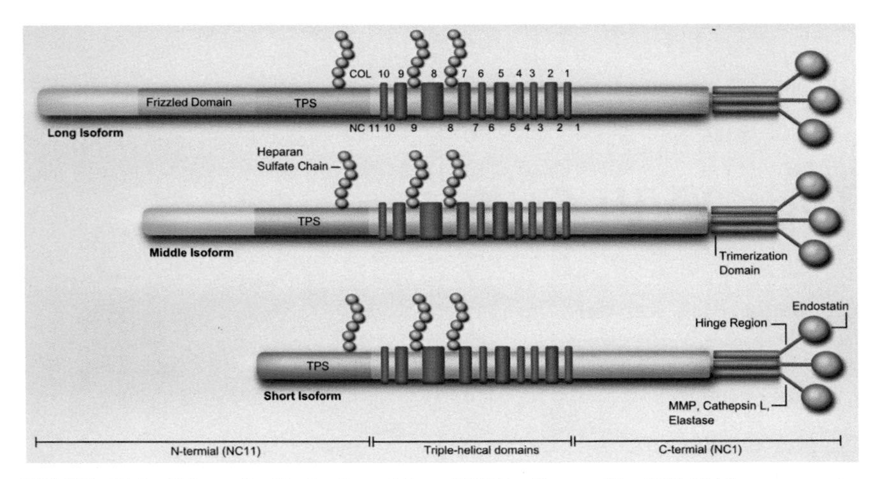

FIGURE 18.1 **Schematic illustration of type XVIII collagen.** The COL18A1 gene encodes three distinct isoforms: short, middle, and long. All three isoforms contain a thrombospondin (TSP) 1–like domain and 10 triple helical collagenous domains (Col1–10) flanked by 11 noncollagenous regions (NC1–11). The NC1 domain contains three functionally distinct subdomains: a trimerization domain, a protease sensitive hinge region, and a C-terminal endostatin domain. The long isoform contains an additional cysteine-rich Frizzled domain.

Using two different promoters and alternative splicing of exon 3, the COL18A1 gene encodes three distinct isoforms: short, middle, and long [3]. The long isoform differs from the middle and short isoforms by containing an additional cysteine-rich Frizzled domain, which when proteolyzed is able to inhibit Wnt/β-catenin signaling by binding to Wnt molecules [4].

All three isoforms contain a thrombospondin (TSP) 1–like domain and 10 triple helical collagenous domains (Col1–10) flanked by 11 noncollagenous regions (NC1–11). The NC1 domain, common to all three isoforms, contains three functionally distinct subdomains: a trimerization domain, a protease-sensitive hinge region, and a C-terminal endostatin domain that has antiangiogenic properties [4] (Fig. 18.1).

Type XVIII collagen also harbors three Ser-Gly consensus sites that function as attachment sites for heparan sulphate glycosaminoglycan (HSGAG) chains [5,6]. Studies show that the HSGAG chains mediate binding with the basement membrane [5]. Furthermore, the HSGAG chains have been reported to interact with L-selectin, an adhesion protein involved in leukocyte migration, and with monocyte chemoattractant protein-1, that through binding with type XVIII collagen, induces α4β1 integrin activation of monocytes [7]. The HSGAG chains also bind the cell adhesion protein receptor protein tyrosine phosphatase σ that is involved in development of the nervous system [8].

All three isoforms of type XVIII collagen are localized in various basement membrane zones where they interact with laminin, perlecan, nidogen, and fibulins [9]. The short isoform is present in vascular and epithelial basement membrane structures, the long isoform is highly expressed in the liver, and the intermediate isoform is not widely expressed [10,11]. Type XVIII collagen has been shown to be located in almost all structures of the human eye, in fat tissue during adipose differentiation, in hair follicles, in articular cartilage, in bone marrow, in the kidneys, and in human plasma [12–17].

Nonsense and missense mutations of type XVIII collagen lead to the autosomal recessive disorder Knobloch syndrome-1, which is characterized by eye abnormalities including high myopia, vitreoretinal degeneration, dislocated lenses, retinal detachment, macular abnormalities, and occipital encephalocele [18]. Lack of either the short or all isoforms of type XVIII collagen causes Knobloch syndrome-1, but patients who lack all isoforms present with more severe eye abnormalities [11].

COL18A1−/− knockout mice likewise suffer from eye abnormalities including abnormal outgrowth of retina vessels, fragile iris, and accumulation of electron-dense deposits [19,20]. Furthermore, COL18A1−/− mice manifest broadened basement membrane structures in the atrioventricular valves of the heart, in kidney tubules, in skin epidermis, and in the choroid plexuses [21]. This suggests that type XVIII collagen not only is involved in eye development but also has an important role in maintaining basement membrane integrity.

Increased amounts of type XVIII collagen and endostatin have been observed in several pathological conditions. Patients with Alzheimer disease show accumulation of the protein in cerebrospinal fluids, amyloid-laden vessels, and senile plaques of the brain, and increased expression of type XVIII collagen has been observed in bullous scleroderma skin [22,23]. Moreover, high serum endostatin levels are associated with increased cancer and cardiovascular death. Type XVIII collagen levels are overexpressed and associated with a poor outcome in patients with non-small-cell lung cancer (NSCLC), colorectal cancer, bladder cancer, and hepatocellular carcinoma [24–27].

ROLE OF ENDOSTATIN IN BIOLOGY AND PATHOLOGY

Endostatin, the carboxyl-terminal fragment of type XVIII collagen, was discovered in 1997 and described as an inhibitor of angiogenesis and tumor growth in murine hemangioendothelioma cells, and has since been intensively studied [28]. The hinge region of endostatin contains several proteolytic cleavage sites where matrix metalloproteinases (MMPs), cathepsins, and elastases are able to cleave off endostatin [29,30]. In its cleaved form, endostatin is able to interact with cell membrane receptors and inhibit endothelial cell proliferation and migration and stimulate cell death by apoptosis [28,31–33]. Through these mechanisms, endostatin has been shown to be an effective inhibitor of tumor growth and has, in a modified recombinant form, been released for treatment of patients with NSCLC in China [34].

The exact mechanism of action of endostatin is not completely known. A broad range of targets have been investigated, and studies point to multiple signaling pathways being involved in mediating the effects of endostatin, as reviewed by Seppinen et al. and Digtyar et al. [6,35]. In brief, endostatin has been shown to directly bind the vascular endothelial growth factor (VEGF) receptor KDR/Flk-1, thereby interfering with VEGF signaling leading to inhibition of proliferation and migration of endothelial cells [32]. Endostatin can also promote β-catenin degradation, leading to cell cycle arrest through inhibition of the cyclin-D1 promoter [36]. Furthermore, studies show that endostatin is able to bind to α5β1 integrin caveolin-1, leading to a decreased cell migration and inhibit the activation or promatrix MMP-2 [37,38].

Endostatin has also been reported to be important in several other pathologies besides cancer. In a mouse ectopic ossification model endostatin was able to reduce bone formation [39]. In other mouse models endostatin has also been observed to inhibit progression of peritoneal sclerosis and diabetic nephropathy [40,41] (Fig. 18.2).

In cardiac diseases such as coronary heart disease and myocardial infarction increased levels of endostatin have been observed [42,43]. Endostatin may suppress aberrant tissue remodeling and scarring [42,44,45]. In a study in rats with myocardial infarction, neutralization of endostatin worsened the outcomes of myocardial infarction [42]. In another study by Yamaguchi et al., a peptide

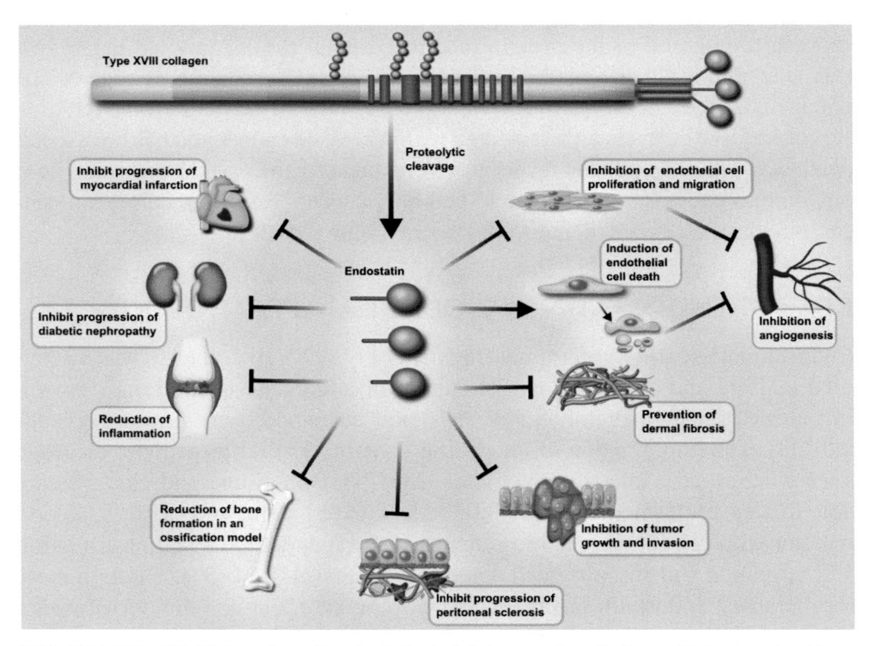

FIGURE 18.2 **Multiple roles of endostatin in biology and pathology.** Endostatin is able to inhibit angiogenesis through induction of endothelial cell death and inhibition of endothelial cell proliferation and migration. In mouse models endostatin have been reported to be able to inhibit the progression of myocardial infarction, inhibit the progression of diabetic nephropathy, reduce inflammation, reduce bone formation, inhibit progression of peritoneal sclerosis and prevent dermal fibrosis.

from the C-terminal end of endostatin prevented transforming growth factor-β–induced dermal fibrosis *ex vivo* and *in vivo* [45].

Endostatin may also have beneficial effects on the development of inflammatory diseases. In a septic mouse model endostatin increased survival by reducing multiple organ dysfunction syndrome, and in rheumatoid arthritis mouse model endostatin was able to reduce inflammation and angiogenesis [46,47] (Fig. 18.2).

BIOMARKERS OF ENDOSTATIN

Endostatin has, in several studies, been shown to have potential as a noninvasive prognostic cancer biomarker, where a high level of endostatin is a marker for worse prognosis [24–27]. Endostatin has also been evaluated as a predictive biomarker and as an efficacy biomarker of cancer treatment, although to a lesser extent [48]. Alba et al. showed that serum endostatin was a predictive factor in metastatic breast cancer patients treated with the aromatase inhibitor letrozole [49]. Öhlund et al. showed that increased serum endostatin levels in patients with pancreatic cancer normalize after surgery or intraperitoneal chemotherapy [50].

Together, these results highlight the potential of endostatin as a cancer biomarkers. However, no biomarkers and antibodies specific to endostatin currently exist. Commercially available antibodies recognize both endostatin still attached to type XVIII collagen and the cleaved and active endostatin fragment. A biomarker specific to the cleaved endostatin fragment may therefore have improved predictive and prognostic value.

Collagen type XVIII	Description	References
Gene name and number	COL18A1 (21q22.3)	RefSeq: NM 130445
Mutations with diseases in humans	Knobloch syndrome	[18]
Null mutation in mice	COL18A1–/– knockout mice suffer from eye abnormalities and manifest broadened basement membrane structures in the atrioventricular valves of the heart, in kidney tubules, in skin epidermis, and in the choroid plexuses	[19–21]
Tissue distribution in healthy states	Localized in various basement membrane zones and is located in almost all structures of the human eye, in fat tissue during adipose differentiation, in hair follicles, articular cartilage, bone marrow, heart, brain, liver, the kidneys, and in human plasma	[12–17]
Tissue distribution in pathologically affected states	Increased amounts of type XVIII collagen and endostatin have been observed in patients with Alzheimer disease, traumatic brain injury, bullous scleroderma skin, NSCLC, colorectal cancer, bladder cancer, and hepatocellular carcinoma	[22–27,51]

Collagen type XVIII	Description	References
Special domains	TSP-1–like domain, Frizzled domain, endostatin, and trimerization domain	[3]
Special neoepitopes	Endostatin, Fizzled domain	[3]
Protein structure	Type XVIII collagen occurs in three N-terminal isoforms: short, middle, and long. Three identical α1 chains assemble into left-handed triple helices. All type XVIII collagen isoforms contain a TSP-1–like domain and 10 triple helical collagenous domains (Col1–10) flanked by 11 noncollagenous regions (NC1–11). The NC1 domain contains a trimerization domain, a protease-sensitive hinge region, and a C-terminal endostatin domain. All three isoforms also contain heparan sulphate side chains, and the long isoform contains a cysteine-rich Frizzled domain	[3]
Binding proteins	Laminins, perlecan, nidogen/entactin, and fibulins	[9]
Known central functions in man	Maintaining basement membrane integrity, inhibition of angiogenesis, inhibition of Wnt/β-catenin signaling and development of the eye	[4,18,21,28]
Animals models with protein affected	NA	
Biomarkers	Endostatin can be used as a prognostic cancer biomarker, but so far no biomarkers specific for the cleaved endostatin fragment exist.	[48]

COL, Collagen; *Col*, collagenous domain; *NA*, not applicable; *NC*, noncollagenous region; *NSCLC*, Non-small-cell lung cancer.

REFERENCES

[1] Pihlajaniemi T, Rehn M. Two new collagen subgroups: membrane-associated collagens and types XV and XVII. Prog Nucleic Acid Res Mol Biol 1995;50:225.

[2] Muragaki Y, Abe N, Ninomiya Y, Olsen BR, Ooshima A. The human alpha 1(XV) collagen chain contains a large amino-terminal non-triple helical domain with a tandem repeat structure and homology to alpha 1(XVIII) collagen. J Biol Chem 1994;269(6):4042.

[3] Elamaa H, Snellman A, Rehn M, Autio-Harmainen H, Pihlajaniemi T. Characterization of the human type XVIII collagen gene and proteolytic processing and tissue location of the variant containing a frizzled motif. Matrix Biol 2003;22(5):427.

[4] Quelard D, Lavergne E, Hendaoui I, Elamaa H, Tiirola U, Heljasvaara R, et al. A cryptic frizzled module in cell surface collagen 18 inhibits Wnt/beta-catenin signaling. PLoS One 2008;3(4):e1878.

[5] Dong S, Cole GJ, Halfter W. Expression of collagen XVIII and localization of its glycosaminoglycan attachment sites. J Biol Chem 2003;278(3):1700.

[6] Seppinen L, Pihlajaniemi T. The multiple functions of collagen XVIII in development and disease. Matrix Biol 2011;30(2):83.

[7] Kawashima H, Watanabe N, Hirose M, Sun X, Atarashi K, Kimura T, et al. Collagen XVIII, a basement membrane heparan sulfate proteoglycan, interacts with L-selectin and monocyte chemoattractant protein-1. J Biol Chem 2003;278(15):13069.

[8] Aricescu AR, McKinnell IW, Halfter W, Stoker AW. Heparan sulfate proteoglycans are ligands for receptor protein tyrosine phosphatase sigma. Mol Cell Biol 2002;22(6):1881.

[9] Kalluri R. Basement membranes: structure, assembly and role in tumour angiogenesis. Nat Rev Cancer 2003;3(6):422.

[10] Saarela J, Rehn M, Oikarinen A, Autio-Harmainen H, Pihlajaniemi T. The short and long forms of type XVIII collagen show clear tissue specificities in their expression and location in basement membrane zones in humans. Am J Pathol 1998;153(2):611.

[11] Suzuki OT, Sertie AL, Der KV, Kok F, Carpenter M, Murray J, et al. Molecular analysis of collagen XVIII reveals novel mutations, presence of a third isoform, and possible genetic heterogeneity in Knobloch syndrome. Am J Hum Genet 2002;71(6):1320.

[12] Bono P, Teerenhovi L, Joensuu H. Elevated serum endostatin is associated with poor outcome in patients with non-Hodgkin lymphoma. Cancer 2003;97(11):2767.

[13] Pufe T, Petersen WJ, Miosge N, Goldring MB, Mentlein R, Varoga DJ, et al. Endostatin/collagen XVIII – an inhibitor of angiogenesis – is expressed in cartilage and fibrocartilage. Matrix Biol 2004;23(5):267.

[14] Sipola A, Nelo K, Hautala T, Ilvesaro J, Tuukkanen J. Endostatin inhibits VEGF-A induced osteoclastic bone resorption in vitro. BMC Musculoskelet Disord 2006;7:56.

[15] Tumbar T, Guasch G, Greco V, Blanpain C, Lowry WE, Rendl M, et al. Defining the epithelial stem cell niche in skin. Science 2004;303(5656):359.

[16] Ylikarppa R, Eklund L, Sormunen R, Muona A, Fukai N, Olsen BR, et al. Double knockout mice reveal a lack of major functional compensation between collagens XV and XVIII. Matrix Biol 2003;22(5):443.

[17] Aikio M, Elamaa H, Vicente D, Izzi V, Kaur I, Seppinen L, et al. Specific collagen XVIII isoforms promote adipose tissue accrual via mechanisms determining adipocyte number and affect fat deposition. Proc Natl Acad Sci USA 2014;111(30):E3043–52.

[18] Sertie AL, Quimby M, Moreira ES, Murray J, Zatz M, Antonarakis SE, et al. A gene which causes severe ocular alterations and occipital encephalocele (Knobloch syndrome) is mapped to 21q22.3. Hum Mol Genet 1996;5(6):843.

[19] Fukai N, Eklund L, Marneros AG, Oh SP, Keene DR, Tamarkin L, et al. Lack of collagen XVIII/endostatin results in eye abnormalities. EMBO J 2002;21(7):1535.

[20] Marneros AG, Olsen BR. Age-dependent iris abnormalities in collagen XVIII/endostatin deficient mice with similarities to human pigment dispersion syndrome. Invest Ophthalmol Vis Sci 2003;44(6):2367.

[21] Utriainen A, Sormunen R, Kettunen M, Carvalhaes LS, Sajanti E, Eklund L, et al. Structurally altered basement membranes and hydrocephalus in a type XVIII collagen deficient mouse line. Hum Mol Genet 2004;13(18):2089.

[22] Santos SN, Oliveira GV, Tavares AL, Massensini AR, Carvalhaes LM, Reljasvaara R, et al. Collagen XVIII and fibronectin involvement in bullous scleroderma. Dermatol Online J 2005;11(1):17.

[23] van HJ, Wilhelmus MM, Heljasvaara R, Pihlajaniemi T, Wesseling P, de Waal RM, et al. Collagen XVIII: a novel heparan sulfate proteoglycan associated with vascular amyloid depositions and senile plaques in Alzheimer's disease brains. Brain Pathol 2002;12(4):456.

[24] Hu TH, Huang CC, Wu CL, Lin PR, Liu SY, Lin JW, et al. Increased endostatin/collagen XVIII expression correlates with elevated VEGF level and poor prognosis in hepatocellular carcinoma. Mod Pathol 2005;18(5):663.

[25] Iizasa T, Chang H, Suzuki M, Otsuji M, Yokoi S, Chiyo M, et al. Overexpression of collagen XVIII is associated with poor outcome and elevated levels of circulating serum endostatin in non-small cell lung cancer. Clin Cancer Res 2004;10(16):5361.
[26] Kantola T, Vayrynen JP, Klintrup K, Makela J, Karppinen SM, Pihlajaniemi T, et al. Serum endostatin levels are elevated in colorectal cancer and correlate with invasion and systemic inflammatory markers. Br J Cancer 2014;111(8):1605.
[27] Szarvas T, Laszlo V, Vom DF, Reis H, Szendroi A, Romics I, et al. Serum endostatin levels correlate with enhanced extracellular matrix degradation and poor patients' prognosis in bladder cancer. Int J Cancer 2012;130(12):2922.
[28] O'Reilly MS, Boehm T, Shing Y, Fukai N, Vasios G, Lane WS, et al. Endostatin: an endogenous inhibitor of angiogenesis and tumor growth. Cell 1997;88(2):277.
[29] Felbor U, Dreier L, Bryant RA, Ploegh HL, Olsen BR, Mothes W. Secreted cathepsin L generates endostatin from collagen XVIII. EMBO J 2000;19(6):1187.
[30] Ferreras M, Felbor U, Lenhard T, Olsen BR, Delaisse J. Generation and degradation of human endostatin proteins by various proteinases. FEBS Lett 2000;486(3):247.
[31] Dhanabal M, Ramchandran R, Waterman MJ, Lu H, Knebelmann B, Segal M, et al. Endostatin induces endothelial cell apoptosis. J Biol Chem 1999;274(17):11721.
[32] Kim YM, Hwang S, Kim YM, Pyun BJ, Kim TY, Lee ST, et al. Endostatin blocks vascular endothelial growth factor-mediated signaling via direct interaction with KDR/Flk-1. J Biol Chem 2002;277(31):27872.
[33] Sudhakar A, Sugimoto H, Yang C, Lively J, Zeisberg M, Kalluri R. Human tumstatin and human endostatin exhibit distinct antiangiogenic activities mediated by alpha v beta 3 and alpha 5 beta 1 integrins. Proc Natl Acad Sci USA 2003;100(8):4766.
[34] Folkman J. Antiangiogenesis in cancer therapy – endostatin and its mechanisms of action. Exp Cell Res 2006;312(5):594.
[35] Digtyar AV, Pozdnyakova NV, Feldman NB, Lutsenko SV, Severin SE. Endostatin: current concepts about its biological role and mechanisms of action. Biochemistry (Mosc) 2007;72(3):235.
[36] Hanai J, Dhanabal M, Karumanchi SA, Albanese C, Waterman M, Chan B, et al. Endostatin causes G1 arrest of endothelial cells through inhibition of cyclin D1. J Biol Chem 2002;277(19):16464.
[37] Kim YM, Jang JW, Lee OH, Yeon J, Choi EY, Kim KW, et al. Endostatin inhibits endothelial and tumor cellular invasion by blocking the activation and catalytic activity of matrix metalloproteinase. Cancer Res 2000;60(19):5410.
[38] Wickstrom SA, Alitalo K, Keski-Oja J. Endostatin associates with integrin alpha5beta1 and caveolin-1, and activates Src via a tyrosyl phosphatase-dependent pathway in human endothelial cells. Cancer Res 2002;62(19):5580.
[39] Sipola A, Ilvesaro J, Birr E, Jalovaara P, Pettersson RF, Stenback F, et al. Endostatin inhibits endochondral ossification. J Gene Med 2007;9(12):1057.
[40] Tanabe K, Maeshima Y, Ichinose K, Kitayama H, Takazawa Y, Hirokoshi K, et al. Endostatin peptide, an inhibitor of angiogenesis, prevents the progression of peritoneal sclerosis in a mouse experimental model. Kidney Int 2007;71(3):227.
[41] Ichinose K, Maeshima Y, Yamamoto Y, Kitayama H, Takazawa Y, Hirokoshi K, et al. Antiangiogenic endostatin peptide ameliorates renal alterations in the early stage of a type 1 diabetic nephropathy model. Diabetes 2005;54(10):2891.
[42] Isobe K, Kuba K, Maejima Y, Suzuki J, Kubota S, Isobe M. Inhibition of endostatin/collagen XVIII deteriorates left ventricular remodeling and heart failure in rat myocardial infarction model. Circ J 2010;74(1):109.

[43] Mitsuma W, Kodama M, Hanawa H, Ito M, Ramadan MM, Hirono S, et al. Serum endostatin in the coronary circulation of patients with coronary heart disease and its relation to coronary collateral formation. Am J Cardiol 2007;99(4):494.

[44] Ren HT, Hu H, Li Y, Jiang HF, Hu XL, Han CM. Endostatin inhibits hypertrophic scarring in a rabbit ear model. J Zhejiang Univ Sci B 2013;14(3):224.

[45] Yamaguchi Y, Takihara T, Chambers RA, Veraldi KL, Larregina AT, Feghali-Bostwick CA. A peptide derived from endostatin ameliorates organ fibrosis. Sci Transl Med 2012;4(136):136–71.

[46] Matsuno H, Yudoh K, Uzuki M, Nakazawa F, Sawai T, Yamaguchi N, et al. Treatment with the angiogenesis inhibitor endostatin: a novel therapy in rheumatoid arthritis. J Rheumatol 2002;29(5):890.

[47] Peng Y, Gao M, Jiang Y, Wang K, Zhang H, Xiao Z, et al. Angiogenesis inhibitor endostatin protects mice with sepsis from multiple organ dysfunction syndrome. Shock 2015;44.

[48] Sund M, Kalluri R. Tumor stroma derived biomarkers in cancer. Cancer Metastasis Rev 2009;28(1–2):177.

[49] Alba E, Llombart A, Ribelles N, Ramos M, Fernandez R, Mayordomo JI, et al. Serum endostatin and bFGF as predictive factors in advanced breast cancer patients treated with letrozole. Clin Transl Oncol 2006;8(3):193.

[50] Öhlund D, Ardnor B, Oman M, Naredi P, Sund M. Expression pattern and circulating levels of endostatin in patients with pancreas cancer. Int J Cancer 2008;122(12):2805.

[51] Mueller CA, Schluesener HJ, Fauser U, Conrad S, Schwab JM. Lesional expression of the endogenous angiogenesis inhibitor endostatin/collagen XVIII following traumatic brain injury (TBI). Exp Neurol 2007;208(2):228.

Chapter 19

Type XIX Collagen

S.H. Nielsen, M.A. Karsdal
Nordic Bioscience, Herlev, Denmark

Chapter Outline

SUMMARY

Type XIX collagen is a member of the fibril-associated collagens with interrupted triple helices, and it acts as a cross-bridge between fibrils and other extracellular molecules. Structurally, type XIX collagen is a homotrimer, composed by three α1 chains. It is expressed in vascular, neuronal, mesenchymal, and epithelial basement membrane zones in the breast, colon, kidney, liver, placenta, prostate, skeletal muscle, skin, and spleen. This collagen has been shown to affect the phenotype for smooth muscle motor dysfunction and hypertension sphincter, by dysregulation of extracellular matrix. The functional role of type XIX collagen is not well established; however, biomarkers could assist the quantification of this protein in both healthy and pathological conditions.

Type XIX collagen is a member of the fibril-associated collagens with interrupted triple helices [1,2] and acts as a cross-bridge between fibrils and other extracellular matrix molecules. Type XIX collagen is a homotrimer of 400 kDa, composed of three α1 chains. Each of these chains consists of 1142 residues. In each, the collagenous region includes five 70–224-residue triple helical subdomains (COL1–COL5) separated by internal 20–44-amino acid noncollagenous domains (NC1–NC6) [1,3]. The gene is localized to 6q12-q14, which is the same region as the α(IX) and α1(XII) collagen genes [1]. It is a minor collagen first discovered in a human rhabdomyosarcoma cell line and now known to be distributed in vascular, neuronal, mesenchymal, and epithelial basement membrane zones (BMZs) in the breast, colon, kidney, liver, placenta, prostate, skeletal muscle, skin, and spleen [3–5]. Collagen type XIX is also expressed by a subset of hippocampal neurons that are necessary for formation of the hippocampal synapses. However, the exact role in the brain is unknown [6].

Biochemistry of Collagens, Laminins and Elastin.

Characterization of the type XIX collagen gene (*Col19a1*) by null mice and structural mutations within the gene, has illustrated the contribution of this collagen to muscle physiology and differentiation of muscle cells [7,8]. The collagen has been shown to affect the phenotype for smooth muscle motor dysfunction and hypertension sphincter. Muscle dysfunction is a cause of a dysregulated matrix. Mice without type XIX collagen exhibit an additional effect known as smooth-to-skeletal muscle cell conversion in the abdominal segment of the esophagus [7]. Besides its role in muscle composition, type XIX collagen is expressed during embryogenesis together with other proteins in extracellular matrix assembly [9]. During invasive stages of breast cancer, collagen type XIX disappears from the BMZ [10]. The inhibitory effect of collagen type XIX was also demonstrated in a murine melanoma model in which the NC1, C-terminal domain of collagen type XIX, inhibited tumor cell migration [5]. This effect is furthermore demonstrated *in vitro* by inhibition of migration capacities of tumor cells [11].

BIOMARKERS OF TYPE XIX COLLAGEN

Assays targeting type XIX collagen have already been developed by using quantitative PCR, Western blot, and enzyme-linked immunosorbent assays in tissues and body fluids [9].

Type XIX collagen	Description	References
Gene name and number	COL19A1, location 6q12q14	NCBI Gene ID: 1310 [1,2]
Mutations with diseases in humans	N.A.	
Null mutation in mice	Col19a null mice display muscle physiology and hypertension sphincter	[7]
Tissue distribution in healthy states	Expressed in endothelial, smooth muscle, neuronal, mesenchymal, and most epithelial cells located in the basement membrane zones in the breast, colon, kidney, liver, placenta, prostate, skeletal muscle, skin, and spleen	[3,4,10]
Tissue distribution in pathologically affected states	Decreased levels of collagen type XIX have been observed in the BMZs in breast cancer at invasive stages	[5,10,11]
Special domains	N.A.	
Special neoepitopes	No special neoepitopes have been discovered but the specificity of MMPs and other proteinases has not yet been determined	[10]
Protein structure	Collagen type XIX is composed by 1142 amino acids, including a 23-amino acid signal peptide, a 268-residue amino-terminal domain, an 832-residue collagenous region, and a 19-amino acid carboxylpeptide. The collagenous domain is composed of five 70-224-residue triple helical subdomains (COL1– COL5) separated by internal 20–44-amino acid noncollagenous domains (NC1–NC6)	[1,3]

Type XIX collagen	Description	References
Binding proteins	N.A.	
Known central functions in man	Acts as a cross-bridge between fibrils and other ECM molecules	[1]
Animals models with protein affected	N.A.	
Biomarkers	N.A.	

COL, Collagen; *NCBI*, National Center for Biotechnology Information; *N.A.*, not applicable; *BMZ*, basement membrane zone; *MMP*, matrix metalloproteinase; *ECM*, extracellular matrix.

REFERENCES

[1] Khaleduzzaman M, Sumiyoshi H, Ueki Y, et al. Structure of the human type XIX collagen (COL19A1) gene, which suggests it has arisen from an ancestor gene of the FACIT family. Genomics 1997;45:304–12.

[2] Gelse K, Poschl E, Aigner T. Collagens–structure, function, and biosynthesis. Adv Drug Deliv Rev 2003;55:1531–46.

[3] Myers JC, Li D, Bageris A, et al. Biochemical and immunohistochemical characterization of human type XIX defines a novel class of basement membrane zone collagens. Am J Pathol 1997;151:1729–40.

[4] Myers JC, Li D, Rubinstein NA, et al. Up-regulation of type XIX collagen in rhabdomyosarcoma cells accompanies myogenic differentiation. Exp Cell Res 1999;253:587–98.

[5] Ramont L, Brassart-Pasco S, Thevenard J, et al. The NC1 domain of type XIX collagen inhibits in vivo melanoma growth. Mol Cancer Ther 2007;6:506–14.

[6] Su J, Gorse K, Ramirez F, et al. Collagen XIX is expressed by interneurons and contributes to the formation of hippocampal synapses. J Comp Neurol 2010;518:229–53.

[7] Sumiyoshi H, Mor N, Lee SY, et al. Esophageal muscle physiology and morphogenesis require assembly of a collagen XIX-rich basement membrane zone. J Cell Biol 2004;166:591–600.

[8] Sumiyoshi H, Laub F, Yoshioka H, et al. Embryonic expression of type XIX collagen is transient and confined to muscle cells. Dev Dyn 2001;220:155–62.

[9] Oudart JB, Brassart-Pasco S, Luczka E, et al. Analytical methods for measuring collagen XIX in human cell cultures, tissue extracts, and biological fluids. Anal Biochem 2013;437:111–7.

[10] Amenta PS, Hadad S, Lee MT, et al. Loss of types XV and XIX collagen precedes basement membrane invasion in ductal carcinoma of the female breast. J Pathol 2003;199:298–308.

[11] Toubal A, Ramont L, Terryn C, et al. The NC1 domain of type XIX collagen inhibits melanoma cell migration. Eur J Dermatol 2010;20:712–8.

Chapter 20

Type XX Collagen

N. Willumsen, M.A. Karsdal
Nordic Bioscience, Herlev, Denmark

Chapter Outline

SUMMARY
Type XX collagen is assigned to the fibril-associated collagens with interrupted triple helices collagen subfamily due to its similarities with types XII and XIV collagen. The tissue distribution is relatively large, and it is elevated in different types of cancer. Structurally, type XX collagen is made up of six fibronectin type III repeat domains, a von Willebrand factor A domain, a laminin G-like domain (also known as the thrombospondin domain), and two triple helical domains. The function of type XX collagen is relatively unknown, and good tools are needed to investigate and quantify the biological role of this collagen.

Type XX collagen, a 136-kDa secreted protein, became a member of the collagen family in 2001 when chick embryo cDNA was isolated and sequenced [1]. Type XX collagen is assigned to the fibril-associated collagens with interrupted triple helices (FACIT) collagen subfamily due to its similarities with types XII and XIV collagen. Structurally, type XX collagen is made up of six fibronectin type III repeat (FN) domains, a von Willebrand factor A (vWA) domain, a laminin G-like (LGL) domain (also known as the thrombospondin domain), and two triple helical domains (Fig. 3, Introduction).

In addition to the canonical sequence, two other isoforms of human type XX collagen produced by alternative splicing have been described [2]. Both of these splice variants contain inserted additional amino acids resulting in masses of 137 and 139 kDa, respectively. Posttranslational modifications have been proposed for type XX collagen. Multiple phosphorylations are found at, or near, the first and fourth FN domains [3]. At the fourth FN domain, a glycosylation at amino acid position 607 has also been found. An acetylation at amino acid position 117 has been described as well.

Type XX collagen forms a protein complex consisting of three α1 chains assembled into left-handed triple helices [4]. In proteomic analysis type XX collagen has been found to colocalize with collagens and other proteins of the extracellular matrix [5]. Based on its similarities to types XII and XIV collagen, it is expected to interact with fibrillar collagens through the C-terminal domain and project the N-terminal domain away from the fibrils [4]. Thus, type XX collagen is able to form bridges linking the fibrillar collagens. Due to the presence of the vWA domain [6], type XX collagen is also likely to be involved in cellular adhesion, migration, differentiation, and signaling.

Initially, type XX collagen mRNA was found most prevalent in corneal epithelium [1]. It was also detectable in embryonic skin, sternal cartilage, and tendon, but it was barely detectable in the calvaria, notochord, or neural retina, suggesting that it was not expressed in these tissues. However, based on a polyclonal antibody (HPA051962) targeting a recombinant protein fragment of type XX collagen, most tissues display moderate-to-strong nuclear and/or cytoplasmic collagen type XX protein staining, with only bile ducts, breast, cerebellum, smooth muscle cells, and soft tissues staining weakly or negatively with this antibody [7]. Using the HPA051962 antibody, several cancer tissues have been found to express type XX collagen as well.

Type XX collagen	**Description**	**References**
Gene name and number	COL20A1 (20q13.33)	Gene ID: 57642
Isoforms (human)	*Isoform 1*: 1284 aa, 136 kDa Canonical sequence *Isoform 2*: 1297 aa, 137 kDa aa 165–166: PA→PGGSEWRET aa 1204–1205: AS→ACESAIQT *Isoform 3*: 1316 aa, 139 kDa aa 1098–1099: R→RGEPGPPGQMGPEGPGGQQGSPGTQGRAVQGPV	[2]
Mutations with diseases in humans	N.A.	
Null mutation in mice	N.A.	
Tissue distribution in healthy states	Most tissues display moderate-to-strong expression with only bile ducts, breast, cerebellum, smooth muscle cells, and soft tissues showing low or negative presence	[7]
Tissue distribution in pathologically affected states	Several cancer tissues	[7]
Special domains	Six FN repeat domains, a vWA domain, an LGL domain/TSP domain, and two triple helical domains	[1,2]

Type XX collagen	Description	References
Special neoepitopes	N.A.	
Protein structure and function	Three α1 chains assemble into left-handed triple helices	[4]
Binding proteins	Fibrillar collagens, similar collagen types XII and XIV	[4]
Known central function	Expected to form bridges linking larger fibrillar collagens, as do types XII and XIV	[4]
	Cellular adhesion, migration, differentiation, and signaling	[6]
Animals models with protein affected	N.A.	
Biomarkers of that protein	IHC antibody (HPA051962)	[7]

COL, Collagen; *aa*, amino acids; *N.A.*, not applicable; *FN*, fibronectin type III repeat; *vWA*, von Willebrand factor A; *LGL*, laminin G-like; *TSP*, thrombospondin; *IHC*, immunohistochemistry.

REFERENCES

[1] Koch M, Foley JE, Hahn R, Zhou P, Burgeson RE, Gerecke DR, et al. Alpha 1(Xx) collagen, a new member of the collagen subfamily, fibril-associated collagens with interrupted triple helices. J Biol Chem 2001;276:23120–6.

[2] Magrane M, Consortium U. UniProt Knowledgebase: a hub of integrated protein data. Database (Oxford) 2011;2011:bar009.

[3] Hornbeck PV, Zhang B, Murray B, Kornhauser JM, Latham V, Skrzypek E. PhosphoSitePlus, 2014: mutations, PTMs and recalibrations. Nucleic Acids Res 2015;43:D512–20.

[4] Ricard-Blum S. The collagen family. Cold Spring Harb Perspect Biol 2011;3:a004978.

[5] Barallobre-Barreiro J, Didangelos A, Schoendube FA, Drozdov I, Yin X, Fernandez-Caggiano M, Willeit P, Puntmann VO, Aldama-Lopez G, Shah AM, et al. Proteomics analysis of cardiac extracellular matrix remodeling in a porcine model of ischemia/reperfusion injury. Circulation 2012;125:789–802.

[6] Colombatti A, Bonaldo P, Doliana R. Type A modules: interacting domains found in several non-fibrillar collagens and in other extracellular matrix proteins. Matrix 1993;13:297–306.

[7] Uhlen M, Bjorling E, Agaton C, Szigyarto CA, Amini B, Andersen E, Andersson AC, Angelidou P, Asplund A, Asplund C, et al. A human protein atlas for normal and cancer tissues based on antibody proteomics. Mol Cell Proteomics 2005;4:1920–32.

Chapter 21

Type XXI Collagen

S.N. Kehlet, M.A. Karsdal
Nordic Bioscience, Herlev, Denmark

Chapter Outline

SUMMARY

Type XXI collagen belongs to the fibril-associated collagens with interrupted triple helices (FACIT) family of collagen proteins. Type XXI collagen may copolymerize with fibrillar collagens via their C-terminal collagenous domains and mediate protein–protein interactions through their N-terminal noncollagenous domains, thereby serving as molecular bridges in the extracellular matrix. Type XXI collagen is expressed in heart, placenta, stomach, jejunum, skeletal muscle, kidney, lung, pancreas, and lymph node. Currently, there are no known mutations in humans and mice null mutation work has not been reported. The biological function of type XXI collagen remains to be revealed; however, data suggest that type XXI collagen may play a role in blood vessel assembly. Biomarkers need to be developed to determine the function and relevance under pathological conditions.

Type XXI collagen belongs to the fibril-associated collagens with interrupted triple helices (FACIT) family of collagen proteins [1]. These collagens consist of two or more collagenous domains that are interrupted by several noncollagenous domains. They often copolymerize with fibrillar collagens via their C-terminal collagenous domains and mediate protein–protein interactions through their N-terminal noncollagenous domains, thereby serving as molecular bridges in the extracellular matrix (ECM) [2,3].

The *COL21A1* open reading frame encodes a 957-amino acid protein with a predicted molecular mass of 99 kDa. The protein possesses a 22-amino acid signal peptide followed by a von Willebrand factor A domain participating in protein–protein interactions and a thrombospondin domain. The C-terminal consists of two collagen triple helical domains interrupted by two short noncollagenous domains [1] (Fig. 3, Introduction). Type XXI collagen mRNA has been shown to be present in ECM-rich tissue, in particular the heart, placenta, stomach, jejunum, skeletal muscle, kidney, lung, pancreas, and lymph node [1]. These tissues are also enriched with type I collagen. Two other members of the FACIT collagen family, type XII and XIV collagens, have been shown to colocalize with type I collagen. Type XXI collagen may also colocalize with type I collagen, suggesting a role of maintaining the integrity of the ECM [1]. However, this hypothesis needs to be investigated further.

The protein is an extracellular component of blood vessel walls and is secreted by vascular smooth muscle cells. It has been speculated that type XXI collagen may have a role in blood vessel assembly since its expression is regulated by platelet-derived growth factor, which induces proliferation and migration of smooth muscle cells [4].

BIOMARKERS OF TYPE XXI COLLAGEN

The ECM can prevent drugs from penetrating into cancer cells [5–7]. A study has shown that *COL21A1* is significantly upregulated in chemoresistant variants of an ovarian cancer cell line [8]. This upregulation indicates that type XXI collagen may be used as a biomarker in ovarian cancer.

Type XXI collagen	Description	References
Gene name and number	COL21A1	NCBI gene ID: 81578
Mutations with diseases in humans	N.A.	
Null mutation in mice	N.A.	
Tissue distribution in healthy states	Heart, placenta, stomach, jejunum, skeletal muscle, kidney, lung, pancreas, and lymph node	[1]
Tissue distribution in pathologically affected states	N.A.	
Special domains	vWA and TSP	[1]
Special neoepitopes	N.A.	
Protein structure and function	Consists of a 22-aa signal peptide, two collagenous domains interrupted by three noncollagenous domains.	[1]
Binding proteins	N.A.	
Known central function	The function remains unknown, but could be implicated in blood vessel assembly	[4]
Animals models with protein affected	N.A.	
Biomarkers	N.A.	

aa, amino acids; *COL*, Collagen; *NCBI*, National Center for Biotechnology Information; *N.A.*, not applicable; *vWA*, von Willebrand factor A; *TSP*, thrombospondin.

REFERENCES

[1] Fitzgerald J, Bateman JF. A new FACIT of the collagen family: COL21A1. FEBS Lett 2001;505:275–80.

[2] Ricard-Blum S, Ruggiero F. The collagen superfamily: from the extracellular matrix to the cell membrane. Pathol Biol (Paris) 2005;53:430–42.

[3] Shaw LM, Olsen BR. FACIT collagens: diverse molecular bridges in extracellular matrices. Trends Biochem Sci 1991;16:191–4.

[4] Chou MY, Li HC. Genomic organization and characterization of the human type XXI collagen (COL21A1) gene. Genomics 2002;79:395–401.

[5] Jain RK. The next frontier of molecular medicine: delivery of therapeutics. Nat Med 1998;4:655–7.

[6] Netti PA, Berk DA, Swartz MA, Grodzinsky AJ, Jain RK. Role of extracellular matrix assembly in interstitial transport in solid tumors. Cancer Res 2000;60:2497–503.

[7] Tannock IF, Lee CM, Tunggal JK, Cowan DS, Egorin MJ. Limited penetration of anticancer drugs through tumor tissue: a potential cause of resistance of solid tumors to chemotherapy. Clin Cancer Res 2002;8:878–84.

[8] Januchowski R, Zawierucha P, Rucinski M, Nowicki M, Zabel M. Extracellular matrix proteins expression profiling in chemoresistant variants of the A2780 ovarian cancer cell line. Biomed Res Int 2014;2014:365867.

Chapter 22

Type XXII Collagen

S.N. Kehlet, M.A. Karsdal
Nordic Bioscience, Herlev, Denmark

Chapter Outline

SUMMARY

Type XXII collagen is a member of the fibril-associated collagen with interrupted triple helices collagen family. The N terminus has a von Willebrand factor A domain and a thrombospondin domain, both containing N-glycosylation sites. Type XXII collagen is predominantly found in the heart and skeletal muscle of human tissue and to a lesser extent in cartilage and skin in mice. It has been identified in specific tissue junctions, namely, the myotendinous junctions (MTJs) that represent the major site of muscle force transmission to tendons. Type XXII collagen binds to integrins, especially $\alpha2\beta1$ and $\alpha11\beta1$, at MTJs, indicating a linking and mechanical stability function of collagen XXII at these points and a role as a cell adhesion ligand. Null mutation results in a muscular dystrophy-like phenotype by disrupting the MTJs. Lastly, the expression level has been shown to be elevated in certain cancer types and associated with kidney function. Currently there are no known biomarkers that may assist in clarifying the functional role of type XXII collagen.

Type XXII collagen is a member of the fibril-associated collagen with interrupted triple helices (FACIT) collagen family. *COL22A1* has an open reading frame of 1626 amino acids, including a 27-amino acid signal peptide. The protein consists of six collagenous domains interrupted by several noncollagenous domains (Fig. 3, Introduction). The N terminus has a von Willebrand factor A (vWA) domain and a thrombospondin (TSP) domain, both containing N-glycosylation sites. Type XXII collagen orthologs have been identified in mouse, zebrafish, and puffer fish [1,2].

mRNA analyses have shown that type XXII collagen is predominantly found in the heart and skeletal muscle of human tissue and to a lesser extent in cartilage and skin in mice. It has been identified in specific tissue junctions, namely,

Biochemistry of Collagens, Laminins and Elastin.

the myotendinous junctions (MTJs) that represent the major site of muscle force transmission to tendons. Muscle cells close to MTJs express type XXII collagen that is deposited in the basement membrane zones of MTJs. Electron microscopy studies have shown that type XXII collagen, unlike other FACIT collagens, does not directly polymerize with fibrillar collagens, but rather associates with components of microfibrils such as fibrillins [1]. The vWA domain may mediate binding of type XXII collagen to components of the basement membrane. It has been shown that type XXII collagen binds to integrins, especially $\alpha2\beta1$ and $\alpha11\beta1$, at MTJs, indicating a linking and mechanical stability function of type XXII collagen at these points and a role as a cell adhesion ligand [3].

Knockdown of *COL22A1* in zebrafish results in a muscular dystrophy–like phenotype. It is accompanied by a highly significant reduction of muscle force caused by a lower level of force transmission capacity together with fiber detachment, all contributing to a reduced number of contracting fibers [4]. This suggests that type XXII collagen functions as a stabilizer and contributes to the maintenance of muscle attachments and efficient transmission of contractile forces. The same study also showed that the dystrophic phenotype could be reversed by microinjection of human recombinant type XXII collagen protein, indicating that *COL22A1* is a candidate gene for muscular dystrophies in humans.

BIOMARKERS OF TYPE XXII COLLAGEN

Preliminary studies suggest that type XXII collagen could have potential as a valuable prognostic biomarker in cancer. mRNA levels have been shown to be significantly higher in head and neck cancer tissue compared to normal tissue from the same individual. The increased mRNA level correlates with lymph node metastasis and pathological stage [5]. Furthermore, a proteomic study has shown that type XXII collagen is a tumor-specific protein characteristic of the colon tumor matrisome [6].

An association between *COL22A1* and the level of serum creatinine, the most important biomarker for assessment of kidney function, has been detected in a meta-analysis of genome-wide data. This could indicate a biological relationship between muscle mass formation and creatinine levels [7].

Type XXII Collagen	Description	References
Gene name and number	COL22A1	NCBI gene ID: 169044
Mutations with diseases in humans	NA	
Null mutation in zebrafish	Knockdown of COL22A1 in zebrafish results in a muscular dystrophy-like phenotype by disrupting the myotendinous junction	[4]
Tissue distribution in healthy states	Myotendinous junctions primarily of the heart and skeletal muscle	[1]

Type XXII Collagen	Description	References
Tissue distribution in pathologically affected states	NA	
Special domains	vWA and TSP	[1]
Special neoepitopes	NA	
Protein structure and function	Comprises a 27-aa signal peptide, six collagenous domains interrupted by several noncollagenous domains	[2]
Binding proteins	Fibrillins, integrins ($\alpha2\beta1$ and $\alpha11\beta1$)	[1,3]
Known central function	The function remains unknown, but may contribute to the mechanical stability of myotendinous junctions	[1,3]
Animals models with protein affected	Zebrafish	[4]
Biomarkers	NA	

aa, amino acids; *COL*, Collagen; *NA*, not applicable; *NCBI*, National Center for Biotechnology Information; *TSP*, thrombospondin; *vWA*, von Willebrand factor A.

REFERENCES

[1] Koch M, Schulze J, Hansen U, Ashwodt T, Keene DR, Brunken WJ, et al. A novel marker of tissue junctions, collagen XXII. J Biol Chem 2004;279:22514–21.

[2] Ricard-Blum S. The collagen family. Cold Spring Harb Perspect Biol 2011;3:a004978.

[3] Zwolanek D, Veit G, Eble JA, Gullberg D, Ruggiero F, Heino J, et al. Collagen XXII binds to collagen-binding integrins via the novel motifs GLQGER and GFKGER. Biochem J 2014;459:217–27.

[4] Charvet B, Guiraud A, Malbouyres M, Zwolanek D, Guillon E, Bretaud S, et al. Knockdown of col22a1 gene in zebrafish induces a muscular dystrophy by disruption of the myotendinous junction. Development 2013;140:4602–13.

[5] Misawa K, Kanazawa T, Imai A, Endo S, Mochizuki D, Fukushima H, et al. Prognostic value of type XXII and XXIV collagen mRNA expression in head and neck cancer patients. Mol Clin Oncol 2014;2:285–91.

[6] Naba A, Clauser KR, Whittaker CA, Carr SA, Tanabe KK, Hynes RO. Extracellular matrix signatures of human primary metastatic colon cancers and their metastases to liver. BMC Cancer 2014;14:518.

[7] Pattaro C, De Grandi A, Vitart V, Hayward C, Franke A, Aulchenko YS, et al. A meta-analysis of genome-wide data from five European isolates reveals an association of COL22A1, SYT1, and GABRR2 with serum creatinine level. BMC Med Genet 2010;11:41.

Chapter 23

Type XXIII Collagen

S.N. Kehlet, M.A. Karsdal
Nordic Bioscience, Herlev, Denmark

Chapter Outline

SUMMARY

Type XXIII collagen is a type II transmembrane collagen. It is found in healthy human and mouse tissues of the lung, cornea, skin, tendon, and amnion and to a lesser extent in the kidney and placenta. Currently, its function is unknown; however, type XXIII collagen is a component of many epithelia and is expressed on epithelial cell surfaces, suggesting a role in the formation or maintenance of cell–cell contacts or in the polarization of epithelial cells. Levels of type XXIII collagen in patient samples have been proposed as a new biomarker of prostate cancer progression and recurrence. Type XXIII collagen has been detected in urine samples of prostate and non-small-cell lung cancer patients, and it is able to discriminate between controls and cancer patients, indicating a potential role as a new diagnostic tool in certain types of cancer. Currently there are no mutations in humans or mice associated with biological functions, and no biomarkers have been put forward.

Type XXIII collagen is a type II transmembrane collagen consisting of an N-terminal noncollagenous domain containing a short cytoplasmic region and a transmembrane stretch followed by extracellular collagenous domains interrupted by noncollagenous sequences [1] (Fig. 3, Introduction). It is found in healthy human and mouse tissues of the lung, cornea, skin, tendon, and amnion and to a lesser extent in the kidney and placenta [2]. Currently the function of type XXIII collagen is unknown. Immunofluorescence analysis using mouse antibodies has revealed that type XXIII collagen is a component of many epithelia and is expressed on epithelial cell surfaces, suggesting a role in the formation or maintenance of cell–cell contacts or in the polarization of epithelial cells. The proteins can either exist as full-length molecules anchored to a cell surface or as shed soluble molecules. Type XXIII collagen is cleaved from the cell surface by furin-mediated proteolysis and releases a 60-kDa soluble domain that forms a

Biochemistry of Collagens, Laminins and Elastin.

multimeric complex [1,2]. At the cell surface of basal keratinocytes, type XXIII collagen directly interacts with integrin $\alpha2\beta1$ and induces adhesion and spreading of keratinocytes [3].

Immunohistochemical analysis has revealed that type XXIII collagen is expressed in a variety of carcinoma tissues including prostate, breast, kidney, lung, and liver tumors [4]. The protein expression is enriched in the highly metastatic rat prostate carcinoma cell line AT6.1 compared with nonmetastatic cell lines [1]. Levels of type XXIII collagen in patient samples have been proposed as a new biomarker of prostate cancer progression and recurrence [5]. A study investigating the potential of type XXIII collagen as a novel lung cancer biomarker also showed that high expression of type XXIII collagen correlates with shorter recurrence-free survival times in non-small-cell lung cancer (NSCLC) patients [4]. This correlation could be explained by the fact that type XXIII collagen facilitates the formation of pulmonary metastases in a mouse metastasis model by mediating initial adhesion to the epithelium, promoting cancer cell aggregation within blood vessels, or both [6].

BIOMARKERS OF TYPE XXIII COLLAGEN

The furin-mediated cleavage of type XXIII collagen may be responsible for the release of protein into biological fluids such as urine and serum. Biomarkers of serum and urine origin have the advantages of being easy to collect, noninvasive, and low cost and they could be used over time to track the course of the disease. Type XXIII collagen has been detected in urine samples of prostate and NSCLC cancer patients and is able to discriminate between controls and cancer patients [4,5], indicating a potential role as a new diagnostic tool in certain types of cancer.

Type XXIII Collagen	Description	References
Gene name and number	COL23A1	NCBI gene ID: 91522
Mutations with diseases in humans	NA	
Null mutation in mice	NA	
Tissue distribution in healthy states	Lung, cornea, skin, tendon, and amnion and to a lesser extent kidney and placenta	[2]
Tissue distribution in pathologically affected states	NA	
Special domains	NA	
Special neoepitopes	Collagen XXIII can be cleaved by furin from the cell surface, generating an extracellular soluble molecule	[1]
Protein structure and function	Consists of an N-terminal cytoplasmic domain, a transmembrane domain, and extracellular triple helical domains interrupted by short noncollagenous domains	[1]

Type XXIII Collagen	Description	References
Binding proteins	Integrin α2β1	[3]
Known central function	The function remains unknown, but may be implicated in the formation or maintenance of cell–cell contacts or in the polarization of epithelial cells	[2]
Animals models with protein affected	NA	
Biomarkers	NA	

COL, Collagen; *NA*, not applicable; *NCBI*, National Center for Biotechnology Information.

REFERENCES

[1] Banyard J, Bao L, Zetter BR. Type XXIII collagen, a new transmembrane collagen identified in metastatic tumor cells. J Biol Chem 2003;278:20989–94.

[2] Koch M, Veit G, Stricker S, Bhatt P, Kutsch S, Zhou P, et al. Expression of type XXIII collagen mRNA and protein. J Biol Chem 2006;281:21546–57.

[3] Veit G, Zwolanek D, Eckes B, Niland S, Kapyla J, Zweers MC, et al. Collagen XXIII, novel ligand for integrin alpha2beta1 in the epidermis. J Biol Chem 2011;286:27804–13.

[4] Spivey KA, Banyard J, Solis LM, Wistuba II, Barletta JA, Gandhi L, et al. Collagen XXIII: a potential biomarker for the detection of primary and recurrent non-small cell lung cancer. Cancer Epidemiol Biomarkers Prev 2010;19:1362–72.

[5] Banyard J, Bao L, Hofer MD, Zurakowski D, Spivey KA, Feldman AS, et al. Collagen XXIII expression is associated with prostate cancer recurrence and distant metastases. Clin Cancer Res 2007;13:2634–42.

[6] Spivey KA, Chung I, Banyard J, Adini I, Feldman HA, Zetter BR. A role for collagen XXIII in cancer cell adhesion, anchorage-independence and metastasis. Oncogene 2012;31:2362–72.

Chapter 24

Type XXIV Collagen

S.H. Nielsen, M.A. Karsdal
Nordic Bioscience, Herlev, Denmark

Chapter Outline

SUMMARY

Type XXIV collagen is a fibrillar collagen containing two collagenous domains and three noncollagenous domains. The amino-terminal peptide contains a thrombospondin N-terminal–like motif and a highly charged segment interspersed with several tyrosine residues. Type XXIV collagen is predominately expressed in the formation of bone; however, it has recently been found in the brain, muscle, kidney, spleen, liver, lung, testis, and ovary, but to a far lesser extent than in bone. *COL24A1* is increased during squamous cell carcinoma of the head and neck and could be a marker of osteoblast differentiation and certain types of cancers. Currently there are no mutations in humans or mice associated with biological functions, and no biomarkers have been put forward.

Type XXIV collagen, encoded by the gene *COL24A1*, is a fibrillar collagen, closely related to types V and XI collagen [1]. Type XXIV collagen is composed of 1714 residues, with two collagenous domains (COL1 and COL2) and three noncollagenous domains (NC1–NC3), flanked by propeptide-like sequences [1–4]. The amino-terminal peptide contains a thrombospondin (TSP) N-terminal domain and a highly charged segment interspersed with several tyrosine residues (Fig. 2, Introduction) [1]. The major triple helix is slightly shorter than other fibrillar collagens (991–997 residues) and contains two glycine substitutions and one G-X-Y imperfection [5,6]. The biological relevance of this imperfection is not yet revealed; however, interruptions in the major triple helical domain of classical fibrillar collagens are known to be a cause of disease [6,7]. The gene is localized to 1p.22.3, and structural considerations and tissue localization suggest a homotrimeric association of type XXIV collagen [1,6].

Type XXIV collagen is predominately expressed in the formation of the mouse embryo as well as in the bone and periosteum of a newborn mouse [2,8].

It is represented together with types I and V collagen in bone, and is known as a marker of osteoblast differentiation and bone formation in mice [8,9]. In humans it was thought to be restricted to bone and cornea together with types I and V collagen [1,7]. However, it has also been found in the brain, muscle, kidney, spleen, liver, lung, testis, and ovary, but to a far lesser extent than in bone [2,3].

The biological role of collagen type XXIV is not revealed [10]. Nevertheless, *Col24a1* mRNA expression profiles have shown that *COL24A1* is increased during squamous cell carcinoma of the head and neck (HNSCC). This collagen could therefore be a useful marker of not only osteoblast differentiation but also prognosis of HNSCC and possibly identify patients at high risk for developing HNSCC [10].

Type XXIV Collagen	Description	References
Gene name and number	COL24A1, location 1p22.3	NCBI gene ID: 255631 [11]
Mutations with diseases in humans	NA	
Null mutation in mice	NA	
Tissue distribution in healthy states	Predominantly in bone, but to a far lesser extent in the brain, muscle, kidney, spleen, liver, lung testis, and ovary	[2,3,8]
Tissue distribution in pathologically affected states	COL24A1 is increased during HNSCC, but the tissue distribution is not fully investigated	[10]
Special domains	TSP N-terminal–like domain	[1]
Special neoepitopes	NA	
Protein structure and function	Collagen type XXIV is a fibrillar collagen closely related to types V and XI collagen, containing two collagenous domains and three noncollagenous domains. The amino-terminal peptide contains a TSP N-terminal-like motif and a highly charged segment interspersed with several tyrosine residues.	[1–3]
Binding proteins	NA	
Known central function	This collagen is known as a marker of osteoblast differentiation and bone formation.	[8]
Animals models with protein affected	NA	
Biomarkers	NA	

COL, collagen; *HNSCC*, squamous cell carcinoma of the head and neck; *NA*, not applicable; *NCBI*, national center for biotechnology information; *TSP*, thrombospondin.

REFERENCES

[1] Koch M, Laub F, Zhou P, et al. Collagen XXIV, a vertebrate fibrillar collagen with structural features of invertebrate collagens: selective expression in developing cornea and bone. J Biol Chem 2003;278:43236–44.

[2] Wang W, Olson D, Liang G, et al. Collagen XXIV (Col24alpha1) promotes osteoblastic differentiation and mineralization through TGF-beta/Smads signaling pathway. Int J Biol Sci 2012;8:1310–22.

[3] Sato K, Yomogida K, Wada T, et al. Type XXVI collagen, a new member of the collagen family, is specifically expressed in the testis and ovary. J Biol Chem 2002;277:37678–84.

[4] Gordon MK, Hahn RA. Collagens. Cell Tissue Res 2010;339:247–57.

[5] Exposito JY, Valcourt U, Cluzel C, et al. The fibrillar collagen family. Int J Mol Sci 2010;11:407–26.

[6] Ricard-Blum S, Ruggiero F. The collagen superfamily. In: Topics in current chemistry. 2005. p. 35–84.

[7] Ricard-Blum S, Ruggiero F. The collagen superfamily: from the extracellular matrix to the cell membrane. Pathol Biol (Paris) 2005;53:430–42.

[8] Matsuo N, Tanaka S, Yoshioka H, et al. Collagen XXIV (Col24a1) gene expression is a specific marker of osteoblast differentiation and bone formation. Connect Tissue Res 2008;49:68–75.

[9] Wada H, Okuyama M, Satoh N, et al. Molecular evolution of fibrillar collagen in chordates, with implications for the evolution of vertebrate skeletons and chordate phylogeny. Evol Dev 2006;8:370–7.

[10] Misawa K, Kanazawa T, Imai A, et al. Prognostic value of type XXII and XXIV collagen mRNA expression in head and neck cancer patients. Mol Clin Oncol 2014;2:285–91.

[11] Boot-Handford RP, Tuckwell DS, Plumb DA, et al. A novel and highly conserved collagen (pro(alpha)1(XXVII)) with a unique expression pattern and unusual molecular characteristics establishes a new clade within the vertebrate fibrillar collagen family. J Biol Chem 2003;278:31067–77.

Chapter 25

Type XXV Collagen

N.G. Kjeld, M.A. Karsdal
Nordic Bioscience, Herlev, Denmark

SUMMARY
Type XXV collagen belongs to the collagen subgroup termed the membrane collagens. Type XXV collagen is primarily expressed in the neurons of the brain and to a lesser extent in the heart, testis, and eye in both humans and mice. Furin convertase is able to release a soluble collagen-like amyloidogenic component (CLAC) by proteolytic cleavage of *COL25A1*. CLAC has been isolated from Alzheimer-diseased brains and identified as a component of senile plaques, which are highly associated with Alzheimer disease. Genetic evidence between the *COL25A1* gene and risk for Alzheimer disease has furthermore been established. *COL25A1* transgenic knockout mice died instantly after birth due to respiratory failure. Further observations showed that muscle motor axons failed to elongate and branch, followed by axon degeneration. These events led to excessive apoptosis, which suggests type XXV collagen is an important molecule. Currently no biomarkers have been put forward.

Type XXV collagen, coded by the gene *COL25A1*, belongs to the collagen subgroup termed the membrane collagens, which also includes collagen types XIII, XVII, and XXIII [1,2].Type XXV collagen is primarily expressed in the neurons of the brain and to a lesser extent in the heart, testis, and eye in both humans and mice [3,4].

The structure of type XXV collagen comprises an N-terminal intracellular noncollagenous domain (NC1), a transmembrane domain, and three extracellular collagenous domains (COL1–3), which are separated by four noncollagenous domains (NC2–4). The configuration of mature type XXV collagen is a triple-helix structure that is stabilized by the collagen-like domains containing repetitive (Gly-X-Y) motifs [1,3,5,6].

Furin convertase is able to release a soluble collagen-like amyloidogenic component (CLAC) by proteolytic cleavage of *COL25A1* [3]. CLAC has been isolated from Alzheimer-diseased brains and identified as a component of senile plaques, which are highly associated with Alzheimer disease [3,5]. CLAC binds to fibrilized amyloid β-peptides, resulting in the formation of partly protease-resistant aggregates [6]. Accumulation of these amyloid β-peptides leads to senile plaque formation. The

Biochemistry of Collagens, Laminins and Elastin.

exact location of the amyloid β-peptide binding motif in CLAC has been identified as the LIKRRLIK sequence located within the NC2 domain [6].

A study in transgenic mice showed that overexpression of type XXV collagen was associated with increased expression of β-site amyloid precursor protein (APP)-cleaving enzyme 1 (BACE-1), which cleaves APP into amyloid β-peptide. Furthermore, accumulation of amyloid β-peptides was observed in the brains of mice, correlating with the increased levels of BACE-1, which indicates that type XXV collagen promotes amyloid plaque formation and contributes to the development of Alzheimer disease [7]. Genetic evidence between the *COL25A1* gene and risk for Alzheimer disease has further been established [8]. Other diseases that are associated with mutations in the *COL25A1* gene are antisocial personality disorder and congenital cranial dysinnervation disorder [3,4,8,9]. *Hb9*-GFP *COL25A1* transgenic knockout mice lacking the furin cleavage site died instantly after birth due to respiratory failure. Further observations showed that muscle motor axons failed to elongate and branch, followed by axon degeneration. These events led to excessive apoptosis, indicating the high importance of CLAC [10].

Type XXV collagen	Description	References
Gene name and number	*COL25A1*	NCBI Gene ID: 84570
Mutations with diseases in humans	An SNP in the *COL25A1* gene is significantly associated with ASPD. Three recessive mutations in type XXV have been identified to be a cause of CCDD. Type XXV collagen is expressed in the brains of patients with Alzheimer disease where the collagen binds to the fibrilized amyloid β-peptides. Genetic evidence indicates an association between the *COL25A1* gene and risk of developing Alzheimer disease.	[3,4,8,9]
Null mutation in mice	*Hb9*-GFP *Col25a1* transgenic knockout mice have been generated by deleting exon 2 of the murine *Col25a1* gene, which contains the furin cleavage site. The mice died instantly after birth due to respiratory failure. Further observations showed that muscle motor axons failed to elongate and branch, followed by axon degeneration. These events led to excessive apoptosis.	[10]
Tissue distribution in healthy states	The expression of type XXV collagen is highly brain specific, being present in the neurons of cerebral neocortices, hippocampus and other subcortical nuclei. Weak expression in the heart, testis and eye in both humans and mice. Type XXV collagen has also been located in conjunctival fibroblasts.	[3,8]

Type XXV collagen	Description	References
Tissue distribution in pathologically affected states	CLAC is distributed in Alzheimer-diseased brains and has been identified as an amyloid plaque component.	[3]
Special domains	Proteolytic cleavage of full type XXV collagen by furin convertase between the 107-KIRIAR sequence and the 117-EAPSE sequence releases CLAC, which contains the extracellular collagen domains that are associated with senile plaques in Alzheimer-diseased brains. The amyloid β-peptide binding motif in CLAC has been located in the IKRRLIK sequence within the NC2 domain.	[3,5,6,11]
Special potential neoepitopes	The furin cleavage site of CLAC is a potential neoepitope.	[3,6]
Protein structure and function	Type XXV collagen is a 654-aa, brain-specific type II transmembrane protein. The structure comprises an intracellular N-terminal noncollagenous domain (NC1), a transmembrane region and three repetitive extracellular Gly-X-Y collagen-like domains (COL1–3) with X and Y being frequently proline and hydroxyproline residues. The COL domains are interrupted by three additional noncollagenous domains (NC2–4). The furin convertase cleavage site is located between Arg-112 and Glu-113, which is able to release the soluble CLAC fragment. The membrane collagens are in general involved in cell–cell and cell–matrix adhesion. CLAC binds specifically to amyloid β-peptides and facilitates their assembly into protease resistant aggregates, thus playing a role in the development of Alzheimer disease.	[3,5–7,10,12]
Binding proteins	Secreted and membrane-bound forms of type XXV collagen bind specifically to the fibrilized forms of amyloid β-peptide.	[3]
Known central function	Type XXV collagen is involved in fibrillization and cell toxicity and protects against proteolysis of amyloid β-peptides in Alzheimer-diseased brains.	[3,11]
Animal models with protein affected	*Hb9*-GFP *Col25a1* transgenic knockout mice. Overexpressed *COL25A1* transgenic mice.	[7,10]
Biomarkers	N.A.	

COL, Collagen; *NCBI*, National Center for Biotechnology Information; *SNP*, single nucleotide polymorphism; *ASPD*, antisocial personality disorder; *CCDD*, congenital cranial dysinnervation disorder; *CLAC*, collagen-like amyloidogenic component; *aa*, amino acid; *N.A.*, not applicable.

REFERENCES

[1] Koch M, Veit G, Stricker S, Bhatt P, Kutsch S, Zhou P, et al. Expression of type XXIII collagen mRNA and protein. J Biol Chem 2006;281:21546–57.

[2] Sertie AL, Sossi V, Camargo AA, Zatz M, Brahe C, Passos-Bueno MR. Collagen XVIII, containing an endogenous inhibitor of angiogenesis and tumor growth, plays a critical role in the maintenance of retinal structure and in neural tube closure (Knobloch syndrome). Hum Mol Genet 2000;9:2051–8.

[3] Hashimoto T, Wakabayashi T, Watanabe A, Kowa H, Hosoda R, Nakamura A, et al. CLAC: a novel Alzheimer amyloid plaque component derived from a transmembrane precursor, CLAC-P/collagen type XXV. EMBO J 2002;21:1524–34.

[4] Shinwari JM, Khan A, Awad S, Shinwari Z, Alaiya A, Alanazi M, et al. Recessive mutations in COL25A1 are a cause of congenital cranial dysinnervation disorder. Am J Hum Genet 2015;96:147–52.

[5] Osada Y, Hashimoto T, Nishimura A, Matsuo Y, Wakabayashi T, Iwatsubo T. CLAC binds to amyloid beta peptides through the positively charged amino acid cluster within the collagenous domain 1 and inhibits formation of amyloid fibrils. J Biol Chem 2005;280:8596–605.

[6] Soderberg L, Kakuyama H, Moller A, Ito A, Winblad B, Tjernberg LO, et al. Characterization of the Alzheimer's disease-associated CLAC protein and identification of an amyloid beta-peptide-binding site. J Biol Chem 2005;280:1007–15.

[7] Tong Y, Xu Y, Scearce-Levie K, Ptacek LJ, Fu YH. COL25A1 triggers and promotes Alzheimer's disease-like pathology in vivo. Neurogenetics 2010;11:41–52.

[8] Forsell C, Bjork BF, Lilius L, Axelman K, Fabre SF, Fratiglioni L, et al. Genetic association to the amyloid plaque associated protein gene COL25A1 in Alzheimer's disease. Neurobiol Aging 2010;31:409–15.

[9] Li D, Zhao H, Kranzler HR, Oslin D, Anton RF, Farrer LA, et al. Association of COL25A1 with comorbid antisocial personality disorder and substance dependence. Biol Psychiatry 2012;71:733–40.

[10] Tanaka T, Wakabayashi T, Oizumi H, Nishio S, Sato T, Harada A, et al. CLAC-P/collagen type XXV is required for the intramuscular innervation of motoneurons during neuromuscular development. J Neurosci 2014;34:1370–9.

[11] Soderberg L, Dahlqvist C, Kakuyama H, Thyberg J, Ito A, Winblad B, et al. Collagenous Alzheimer amyloid plaque component assembles amyloid fibrils into protease resistant aggregates. FEBS J 2005;272:2231–6.

[12] Ricard-Blum S, Ruggiero F. The collagen superfamily: from the extracellular matrix to the cell membrane. Pathol Biol (Paris) 2005;53:430–42.

Chapter 26

Type XXVI Collagen

N.G. Kjeld, M.A. Karsdal
Nordic Bioscience, Herlev, Denmark

SUMMARY
Type XXVI collagen contains collagenous domains in its structure, but does not properly fit within any of the collagen family subgroups. Trimers of type XXVI collagen are formed from the intermolecular disulfide bonds between the NC1 regions. Such trimerization formation is a special characteristic of type XXVI collagen compared with fibrillar collagens, where trimer formation is associated with the C-propeptide region. Type XXVI collagen is expressed in the testis and ovary of adult tissues. The highest levels are found in the reproductive tissues of neonates. Consequently, type XXVI collagen has been suggested to be associated with generation and modeling of tissues. The research on type XXVI collagen is only now emerging, and no biomarkers or mutations associated with this protein have been discovered.

Type XXVI collagen, encoded by the gene *COL26A1*, is a 441-amino acid protein encoded by the emilin/multimerin domain containing protein 2 on chromosome 7q22.1 [1]. Type XXVI collagen contains collagenous domains in its structure, but does not properly fit within any of the collagen family subgroups.

The structure of type XXVI collagen comprises two collagenous domains (COL1 and COL2) consisting of Gly-X-Y repeats and three noncollagenous domains (NC1–3). It has been shown that type XXVI collagen forms trimers, accumulates in the extracellular matrix of mouse tissues and is secreted in vitro [2].

The NC1 domain contains a putative endoplasmic reticulum–targeting signal sequence and an emilin domain, which consists of 13 cysteine residues [2,3]. Type XXVI collagen trimers are formed from the intermolecular disulfide bonds between the NC1 regions. Such trimerization formation is a special characteristic of type XXVI collagen compared with fibrillar collagens, where trimer formation is associated with the C-propeptide region [2]. NC1 contains a furin protease cleavage site (RRRR) that is also present in the transmembrane collagens such as types XIII and XXV collagen. The cleavage site is located just before the COL1 domain, but it does not undergo furin cleavage as do the transmembrane proteins.

Biochemistry of Collagens, Laminins and Elastin.

Type XXVI collagen is expressed in the testis and ovary of adult tissues. The highest levels are found in the reproductive tissues of neonates. It has been shown that during early testis and ovary development, type XXVI collagen is highly expressed in myoid cells and pretheca cells in the testis and ovary, respectively, causing accumulation of type XXVI collagen in these tissues [2]. The functional role of type XXVI collagen has therefore been suggested to be in the early development of testis and ovary as an extracellular matrix component [2].

Type XXVI collagen	Description	References
Gene name and number	*COL26A1*	NCBI Gene ID: 136227
Mutations with diseases in humans	N.A.	
Null mutation in mice	N.A.	
Tissue distribution in healthy states	Type XXVI collagen is expressed in the testis and ovary in adult tissues. Highest levels are found in the reproductive tissues of neonates	[2]
Tissue distribution in pathological affected states	N.A.	
Special domains	The NC1 domain contains a putative ER-targeting signal sequence, and an EMI domain, which is made up by 13 cysteine residues. The NC1 domain also contains a putative furin protease cleavage site (RRRR)	[2,3]
Special neoepitopes	N.A.	[2]
Protein structure and function	The structure comprises two collagenous domains (COL1 and COL2) consisting of Gly-X-Y repeats and three noncollagenous domains (NC1–3). The NC1 domain contains a putative ER-targeting signal sequence, an EMI domain and a furin protease cleavage site. It has been shown that type XXVI collagen forms trimers, accumulates in the extracellular matrix of mouse tissues and is secreted in vitro	[2,3]
Binding proteins	HSP47, a collagen-specific molecular chaperone that binds to the triple helical domain	[2]
Known central function	Type XXVI collagen has been suggested to be involved in the early development of testis and ovary as an extracellular matrix component	[2]
Animal models with protein affected	N.A.	
Biomarkers	N.A.	

COL, Collagen; *NCBI*, National Center for Biotechnology Information; *ER*, endoplasmic reticulum; *EMI*, emilin; *N.A.*, not applicable.

REFERENCES

[1] Pasaje CF, Kim JH, Park BL, Cheong HS, Kim MK, Choi IS, et al. A possible association of EMID2 polymorphisms with aspirin hypersensitivity in asthma. Immunogenetics 2011; 63:13–21.

[2] Sato K, Yomogida K, Wada T, Yorihuzi T, Nishimune Y, Hosokawa N, et al. Type XXVI collagen, a new member of the collagen family, is specifically expressed in the testis and ovary. J Biol Chem 2002;277:37678–84.

[3] Leimeister C, Steidl C, Schumacher N, Erhard S, Gessler M. Developmental expression and biochemical characterization of Emu family members. Dev Biol 2002;249:204–18.

Chapter 27

Type XXVII Collagen

F. Genovese, M.A. Karsdal
Nordic Bioscience, Herlev, Denmark

Chapter Outline

SUMMARY
Type XXVII collagen is a fibrillar collagen; however, the molecular structure is different from that of other fibrillar collagens: type XXVII collagen forms characteristic nonstriated fibrils, its triple helix domain is shorter than that of other fibrillar collagens and it presents two interruptions, and it lacks the characteristic N-terminal telopeptide-like region and the minor triple helical domain. This suggests that the function of this fibrillar collagen is different from that of other fibrillar collagens. In adults, type XXVII collagen is primarily expressed in cartilage. During development, the expression is particularly high in developing bones, and mutations in mice lead to chondrodysplasia. In humans the only disease associated with certainty with mutations in the type XXVII collagen gene is Steel syndrome, which is characterized by dislocated hips, prominent forehead and a long oval-shaped face, short stature, scoliosis, and other skeletal abnormalities.

Type XXVII collagen is a fibrillar collagen encoded by the gene *COL27A1*, which was first identified by Pace et al. in 2003 [1]. The human gene is localized on chromosome 9q32, whereas the mouse homologous gene is located on chromosome 4. Its sequence is particularly conserved in human, mouse, and fish. Most of the research done so far on type XXVII collagen has been performed in mouse or zebrafish [2–6].

As with other fibrillar collagens, type XXVII collagen contains an N-terminal and a C-terminal propeptide. The full-length procollagen is 1860 amino acids long, consisting of a 41-amino acid predicted signal peptide, the N-terminal propeptide (583 amino acids) that contains a thrombospondin (TSP) domain, the collagenous triple-helix domain (994 amino acids), and the C-terminal propeptide (242 amino acids) [1]. Together with type XXIV collagen, the type XXVII collagen gene has been defined as belonging to a third class (type C) within the family of fibrillar collagens [7,8]. Its molecular structure is different from that of other fibrillar collagens because type XXVII collagen forms characteristic nonstriated fibrils whose width has been observed to vary from 10 to 80 nm according to the preparation method used [6,9]. The type XXVII collagen triple-helix domain is shorter than those in the other fibrillar collagens, and it presents two interruptions. Moreover, it lacks the fibrillar collagen characteristic N-terminal telopeptide-like region and the minor triple helical domain [6,9].

As with types II and XI collagen, type XXVII collagen is primarily expressed in cartilage in adult organisms. Its expression is regulated by factors SOX9 and Lc-Maf in chondrocytes [5,10]. During embryonic and fetal development, the expression of COL27A1 in mice is more broadly localized in the anlagen of endochondral bone [6]; in the developing lungs, ears, and colon [1]; in the retina and in the cornea of the eye; and in the major arteries of the heart [9]. This suggests that type XXVII collagen plays a key role in the developmental phases [9]. Its role in cartilage during fetal development has been studied. COL27A1 is expressed at 18–20 weeks of gestation in human fetal epiphyseal cartilage [11]. Type XXVII collagen is mainly deposited in skeletal tissues at sites of transition from cartilage to bone. In developing endochondral bone, type XXVII collagen plays a role in the transition of cartilage to bone during skeletogenesis [6]. Its expression is particularly high in the matrix surrounding proliferative chondrocytes in the epiphyseal growth plate [4]. To better understand the role of type XXVII collagen, deletion and mutation knockin mice have been generated [4]. Mice carrying a heterozygous 87-amino acid deletion in the collagenous domain survived into adulthood, but with growth impairments, whereas homozygous mice showed skeletal abnormalities, severe chondrodysplasia, and died at birth due to respiratory problems [4]. Mutant experiments in zebrafish have shown that knockdown in zebrafish embryos of COL27A1A alone and in combination with COL27A1B resulted in delayed and decreased vertebral mineralization, morphological abnormalities, and scoliosis [2].

In humans the only disease associated with certainty with mutations in the type XXVII collagen gene is Steel syndrome, characterized by dislocated hips, prominent forehead and a long oval-shaped face, short stature, scoliosis, and other skeletal abnormalities [12]. A mutation responsible for the syndrome has been identified in the homozygous missense variant Gly697Arg [12]. Moreover, it has been suggested that mutations influencing the structure and properties of tenascin-C and type XXVII collagen could possibly lead to a risk for developing Achilles tendinopathy [13]. Preliminary evidence has been presented showing that single-nucleotide polymorphisms in COL27A1 can be involved in the development of Tourette syndrome [14].

Type XXVII Collagen	**Description**	**References**
Gene name and number	COL27A1, located on chromosome 9q32	NCBI gene ID: 85301
Mutations with diseases in humans	Steel syndrome; Tourette syndrome; Achilles tendinopathy	[12–14]
Mutations in mice	Homozygous deletion of 87 aa in the collagenous domain. Skeletal abnormalities, severe chondrodysplasia, and death at birth.	[4]
Tissue distribution in healthy states	Adults: cartilage. Children and adolescents: epiphyseal growth plate. During embryonic and fetal development: endochondral bone, lungs, ear, colon, retina, cornea, and major arteries of the heart.	[1,6,9,11]
Tissue distribution in pathological affected states	NA	
Special domains	TSP domain in the N-terminal propeptide	[1]
Special neoepitopes	NA	
Protein structure and function	N-terminal propeptide (583 aa), collagenous triple-helix domain (994 aa), and C-terminal propeptide (242 aa).	
Binding proteins	NA	
Known central function	Type XXVII collagen may play a role during endochondral bone formation, including calcification and degradation of cartilage	[1]
Animal models with protein affected	Knockdown of COL27A1A alone and in combination with COL27A1B in zebrafish embryos: delayed and decreased vertebral mineralization, morphological abnormalities, and scoliosis.	[2]
Biomarkers of that protein	NA	

aa, amino acid; *COL*, collagen; *NA*, not applicable; *NCBI*, national center for biotechnology information; *TSP*, thrombospondin.

REFERENCES

[1] Pace JM, Corrado M, Missero C, Byers PH. Identification, characterization and expression analysis of a new fibrillar collagen gene, COL27A1. Matrix Biol 2003;22:3–14.

[2] Christiansen HE, Lang MR, Pace JM, Parichy DM. Critical early roles for col27a1a and col27a1b in zebrafish notochord morphogenesis, vertebral mineralization and post-embryonic axial growth. PLoS One 2009;4:e8481.

[3] Duran I, Csukasi F, Taylor SP, Krakow D, Becerra J, Bombarely A, et al. Collagen duplicate genes of bone and cartilage participate during regeneration of zebrafish fin skeleton. Gene Expr Patterns 2015;19(1–2):60–9.

[4] Plumb DA, Ferrara L, Torbica T, Knowles L, Mironov Jr A, Kadler KE, et al. Collagen XXVII organises the pericellular matrix in the growth plate. PLoS One 2011;6:e29422.

[5] Mayo JL, Holden DN, Barrow JR, Bridgewater LC. The transcription factor Lc-Maf participates in Col27a1 regulation during chondrocyte maturation. Exp Cell Res 2009;315:2293–300.
[6] Hjorten R, Hansen U, Underwood RA, Telfer HE, Fernandes RJ, Krakow D, et al. Type XXVII collagen at the transition of cartilage to bone during skeletogenesis. Bone 2007;41:535–42.
[7] Boot-Handford RP, Tuckwell DS, Plumb DA, Rock CF, Poulsom R. A novel and highly conserved collagen (pro(alpha)1(XXVII)) with a unique expression pattern and unusual molecular characteristics establishes a new clade within the vertebrate fibrillar collagen family. J Biol Chem 2003;278:31067–77.
[8] Wada H, Okuyama M, Satoh N, Zhang S. Molecular evolution of fibrillar collagen in chordates, with implications for the evolution of vertebrate skeletons and chordate phylogeny. Evol Dev 2006;8:370–7.
[9] Plumb DA, Dhir V, Mironov A, Ferrara L, Poulsom R, Kadler KE, et al. Collagen XXVII is developmentally regulated and forms thin fibrillar structures distinct from those of classical vertebrate fibrillar collagens. J Biol Chem 2007;282:12791–5.
[10] Jenkins E, Moss JB, Pace JM, Bridgewater LC. The new collagen gene COL27A1 contains SOX9-responsive enhancer elements. Matrix Biol 2005;24:177–84.
[11] Pogue R, Sebald E, King L, Kronstadt E, Krakow D, Cohn DH. A transcriptional profile of human fetal cartilage. Matrix Biol 2004;23:299–307.
[12] Gonzaga-Jauregui C, Gamble CN, Yuan B, Penney S, Jhangiani S, Muzny DM, et al. Mutations in COL27A1 cause Steel syndrome and suggest a founder mutation effect in the Puerto Rican population. Eur J Hum Genet 2015;23:342–6.
[13] Saunders CJ, van der Merwe L, Posthumus M, Cook J, Handley CJ, Collins M, et al. Investigation of variants within the COL27A1 and TNC genes and Achilles tendinopathy in two populations. J Orthop Res 2013;31:632–7.
[14] Liu S, Yu X, Xu Q, Cui J, Yi M, Zhang X, et al. Support of positive association in family-based genetic analysis between COL27A1 and Tourette syndrome. Sci Rep 2015;5:12687.

Chapter 28

Type XXVIII Collagen

A. Arvanitidis, M.A. Karsdal
Nordic Bioscience, Herlev, Denmark

Chapter Outline

SUMMARY

The human and murine type XVVIII collagen structurally resembles collagen type VI. Type XXVIII collagen is mainly located in peripheral nerves surrounding most non-myelinating glial cells and dorsal root ganglia, but it is also present in skin calvaria, at the nodes of Ranvier, and as a component of the peripheral nervous system nodal gap. Interestingly, type XXVIII collagen is overexpressed in bleomycin-induced lung and in mouse hepatocarcinoma, suggesting it is involved in damage repair processes. Type XXVIII collagen has been located in murine lungs associated with basement membranes, albeit the expression is not restricted to this membrane. The research on type XXVIII collagen is just emerging. Currently there are no mutations in humans or mice associated with biological functions, and no biomarkers have been put forward.

The *COL28A1* gene on human chromosome 7p21.3 and on mouse chromosome 6A1 encodes collagen XXVIII with very restricted expression on tissues. Type XXVIII collagen is mainly located in peripheral nerves and surrounds all nonmyelinating glial cells (with the exception of type II terminal Schwann cells in the hairy skin) and dorsal root ganglia. It is also present in skin and calvaria [1–4] and can be found at the nodes of Ranvier and is a component of the peripheral nervous system nodal gap [1].

Type XXVIII collagen mRNA is expressed in newborn sciatic nerves, but not in adults, although the protein persists into adulthood, suggesting a long half-life for the molecule [1,5]. Col28A1 was weakly expressed in the frontal cortex, heart, liver, adrenal gland, gallbladder, pancreas, esophagus, colon, rectum, urinary bladder, and prostate of adults [6].

Type XXVIII collagen has been also detected in the murine lung associated with basement membranes around vessels, airways, and alveoli [7]. Furthermore, in zebrafish studies, type XXVIII collagen (Col28a1a and Col28a2b) is mainly expressed in the skeletal muscle and in the skin, making it likely to be implicated during the developmental stage. Lastly, there is a unique expression of Col28a1a and Col28a2b in the liver and thymus that suggests the need for further analysis as the expression in mammals can be broader than acknowledged at present [1].

The human and murine type XVVIII collagen structurally resembles type VI collagen. It consists of two von Willebrand factor A (vWA) domains flanking a 528-amino acid collagenous domain. There are 16 uniformly arranged, very short imperfections (G1G and G4G) in the Gly-X-Y repeat, which is unique among the known collagens [3,8]. The triple helical part is longer than that in type VI collagen. The C terminus comprises a unique domain, nonhomologous to previously known collagen domains, and a Kunitz family serine protease inhibitor module similar to that in the vWA domain. Type XXVIII collagen forms dimers as shown by electrophoresis in composite polyacrylamide/agarose gels under nonreducing conditions [2,5].

Type XXVIII collagen is particularly associated with unmyelinated fibers, with the nodes of Ranvier of myelinated axons and with terminal Schwann cells in sensory end organs. Worth noting is the enhanced expression of type XXVIII collagen in a mouse model of Charcot–Marie–Tooth disease, a disease that is characterized by dysmyelinated nerve fibers, further confirming the correlation between absence of myelin and presence of type XXVIII collagen [1,3]. Type XXVIII collagen thus shows promise as a marker of nonmyelinated zones of the peripheral somatosensory system.

Type XXVIII collagen was found in very low levels in healthy lung tissue, but it was overexpressed in bleomycin-induced lung injury, defined as a strong staining with a patchy appearance in fibrotic foci that did not colocalize with α-smooth muscle action–positive myofibroblasts. This indicates that a subset of cells specifically expressing type XXVIII collagen can be involved in the tissue repair process [7]. In addition, upregulation of Col28a1 was also observed in mouse hepatocarcinoma [6], which could point to a role of Col28a1 in the fibrotic stage of this disease.

Type XXVIII collagen	Description	References
Gene name and number	COL28A1	
Mutations with diseases in humans	N.A.	
Null mutation in mice	N.A.	
Tissue distribution in healthy states	Mainly located in peripheral nerves surrounding most nonmyelinating glial cells and dorsal root ganglia. Also present in skin calvaria, at the nodes of Ranvier. Component of the PNS nodal gap	[1–4]
Tissue distribution in pathological affected states	Overexpressed in bleomycin-induced lung and in mouse hepatocarcinoma	[6,7]
Special domains	None	
Special neoepitopes	None known	

Type XXVIII collagen	Description	References
Protein structure and function	Homotrimeric molecule with its α chain composed of a 528-amino acid collagenous domain flanked by two vWA modules, which are known to be involved in protein–protein interactions. The C terminus consists of a module without homology to any known domain, followed by a Kunitz family serine protease inhibitor module	[3]
Binding proteins	Forms dimers	[1]
Known central function	Structural and functional role in the PNS	[3]
Animals models with protein affected	N.A.	
Biomarkers of that protein	N.A.	

COL, Collagen; *N.A.*, not applicable; *PNS*, peripheral nervous system; *vWA*, von Willebrand factor A.

REFERENCES

[1] Veit G, Kobbe B, Keene DR, Paulsson M, Koch M, Wagener R. Collagen XXVIII, a novel von Willebrand factor A domain-containing protein with many imperfections in the collagenous domain. J Biol Chem 2006;281:3494–504.

[2] Gebauer JM, Kobbe B, Paulsson M, Wagener R. Structure, evolution and expression of collagen XXVIII: lessons from the zebrafish. Matrix Biol 2015;49.

[3] Grimal S, Puech S, Wagener R, Venteo S, Carroll P, Fichard-Carroll A. Collagen XXVIII is a distinctive component of the peripheral nervous system nodes of ranvier and surrounds nonmyelinating glial cells. Glia 2010;58:1977–87.

[4] Griffin JW, Thompson WJ. Biology and pathology of nonmyelinating Schwann cells. Glia 2008;56:1518–31.

[5] Gordon MK, Hahn RA. Collagens. Cell Tissue Res 2010;339:247–57.

[6] Lai KK, Shang S, Lohia N, Booth GC, Masse DJ, Fausto N, et al. Extracellular matrix dynamics in hepatocarcinogenesis: a comparative proteomics study of PDGFC transgenic and Pten null mouse models. PLoS Genet 2011;7:e1002147.

[7] Schiller HB, Fernandez IE, Burgstaller G, Schaab C, Scheltema RA, Schwarzmayr T, et al. Time- and compartment-resolved proteome profiling of the extracellular niche in lung injury and repair. Mol Syst Biol 2015;11:819.

[8] Thiagarajan G, Li Y, Mohs A, Strafaci C, Popiel M, Baum J, et al. Common interruptions in the repeating tripeptide sequence of non-fibrillar collagens: sequence analysis and structural studies on triple-helix peptide models. J Mol Biol 2008;376:736–48.

Chapter 29

Laminins

D. Guldager Kring Rasmussen, M.A. Karsdal
Nordic Bioscience, Herlev, Denmark

Chapter Outline

SUMMARY

Laminins are a major constituent of the basement membrane which is an intricate meshwork of proteins separating the epithelium, mesothelium, and endothelium from connective tissue. There are 15 different laminins, each consisting of a unique combination of three subchains. The combination of chains confers some tissue specificity. Laminins are essential for the function of the basement membrane as most null mutations are lethal. Just as collagens, laminins are structural proteins with a helical region formed by heptads. The heptad has a less strict organization compared to collagens, where the helical domain consists of a strict building block of three amino acids. Laminins carry out a central role in organizing the complex interactions of the basement membranes. This is seen through the wide range of interaction partners such as syndecan, nidogen, collagen, integrins, dystroglycan, and heparin. Mutations in different laminin chains are associated with diseases such as Alport syndrome, epidermolysis bullosa, and muscle dystrophies. The laminin family is large, and biomarkers for measuring laminins are emerging. The essential role of laminins in the basement membrane can be summarized by the following: Even the most exceptional construction will not last on a poor foundation.

Biochemistry of Collagens, Laminins and Elastin.

Laminins are a major constituent of the basement membrane, a thin layer of specialized extracellular matrix (ECM) that surrounds epithelia, endothelia, muscle cells, fat cells, Schwann cells, peripheral nerves, and the entire central nervous system. The basement membrane is an intricate meshwork composed of laminins, collagen IV, nidogens, and sulfated proteoglycans [1,2]. Even though the basement membrane consists of the same proteins, different isoforms of these combine to form a structurally and functionally diverse basement membrane.

Laminins are multidomain proteins and are integral parts in the matrix of the basement membrane. During early development of mice embryos, the presence of laminin is sufficient for the formation of basement membrane–like structures in the absence of the other major basement membrane protein, type IV collagen [3]. Structurally, laminins can generally be viewed as intertwined, heterotrimeric glycoproteins of one of three types: cross shaped, Y shaped, or rod shaped [4]. Each laminin is made up of an α, β, and γ chain (Table 29.1). The formation of laminin trimers seems to depend on the C-terminal portions of their coiled regions [5]. The difference in structure of the heterotrimeric laminin glycoproteins is mainly through differences in the α chains that give rise to the three structural versions mentioned above. One unique feature of all laminin α chains is that they contain a large C-terminal globular (G) domain with five modules known as laminin G (LG) domains [6–8]. Currently, five distinct α, three β, and three γ laminin chains have been identified, and together they form at least 15 isoforms of which 15 are presented in the following sections (see Table 29.1) [9,10]. Contradicting evidence for the existence of the laminin β4 chain has been presented, and it remains controversial whether it has been found in vivo, so this chain is not discussed in this chapter.

During development, different laminins are present in various tissues. As part of the maturation of organs, these early laminins are often replaced by other laminins. This phenomenon ultimately produces tissue-specific laminins that often depend on the type of laminin α chain. One laminin that defies this general pattern is laminin-8 (α4β1γ1) that is expressed by all endothelial cells regardless of their stage of development [11].

A wide range of proteins interact with laminins. Among these proteins are cell surface receptors. The cell surface receptors that interact with laminins can be separated into integrins and nonintegrins. For example, integrins that have been shown to interact with laminins-10 and -11 include α3β1, α6β1, and α6β4 [12]. Currently, three different classes of nonintegrin cell surface receptors are known. Two of these classes, dystroglycan and syndecan, are proteoglycans, and among the latter class, Lutheran blood group glycoprotein (also known as basal cell adhesion molecule), is a member of the immunoglobulin superfamily. The extracellular α chain of dystroglycan primarily binds to the G domains of laminin α1 and α2, and with less affinity for other G domains of other α chains [13–15]. Interactions of the specific laminin chains can be seen in the tables in the following sections.

In an attempt to increase the clarity of the chain composition of laminins, Aumailley et al. [16] came up with a new nomenclature that is being increasingly

TABLE 29.1 Laminin Chain Composition, Simplified Nomenclature, and Sites of Expression

Protein	Chains	Alternative Nomenclature	Major Site(s) of Expression
Laminin-1	$\alpha1\beta1\gamma1$	Laminin-111	Developing epithelia
Laminin-2	$\alpha2\beta1\gamma1$	Laminin-211	All myogenic tissues, peripheral nerves, mesangial matrix of glomerulus
Laminin-3	$\alpha1\beta2\gamma1$	Laminin-121	Myotendinous junctions
Laminin-4	$\alpha2\beta2\gamma1$	Laminin-221	Neuromuscular junction, mesangial matrix of glomerulus
Laminin-5	$\alpha3\beta3\gamma2$	Laminin-332	Epidermis
Laminin-6	$\alpha3\beta1\gamma1$	Laminin-311	Epidermis
Laminin-7	$\alpha3\beta2\gamma1$	Laminin-321	Epidermis
Laminin-8	$\alpha4\beta1\gamma1$	Laminin-411	Endothelium, adipose tissue, peripheral nerves
Laminin-9	$\alpha4\beta2\gamma1$	Laminin-421	Endothelium, smooth muscle, neuromuscular junction, mesangial matrix of glomerulus
Laminin-10	$\alpha5\beta1\gamma1$	Laminin-511	Mature epithelium, mature endothelium, smooth muscle, peripheral nerves
Laminin-11	$\alpha5\beta2\gamma1$	Laminin-521	Mature epithelium, mature endothelium, smooth muscle, neuromuscular junction, glomerular basement membrane
Laminin-12	$\alpha2\beta1\gamma3$	Laminin-213	Surface of ciliated epithelia
Laminin-13	$\alpha3\beta2\gamma3$	Laminin-323	Central nervous system, retina
Laminin-14	$\alpha4\beta2\gamma3$	Laminin-423	Central nervous system, retina
Laminin-15	$\alpha5\beta2\gamma3$	Laminin-523	Central nervous system, retina

used in the laminin field. In this new nomenclature, laminin-1 would be called laminin-111, which is more explanatory than the original name because it reveals the composition of chains (see Table 29.1).

In the following sections, select findings are presented on each of the laminins.

LAMININ-1

Laminins-1 and -10 seem to be the only two laminins expressed in early postimplantation embryos. Laminin-1 is the core laminin in the Reichert membrane, an extraembryonic basement membrane [17]. Because the laminin α1 chain interacts with dystroglycan and the absence of either leads to early lethality during development, both seem to be essential for structural integrity of the Reichert membrane. The importance of laminin composition in basement membranes was evaluated by Miner et al. [17], who showed that the laminin α1, β1, and γ1 chains were necessary for development. In mice lacking either of these chains, the embryos died early and no gastrulation occurred [17]. The phenotype of the mice could partially be rescued by overexpression of the laminin α5 chain through a transgene, but the Reichert membrane and blood sinuses were not formed [17]. An in vitro study showed that laminin-1 was able to cause differentiation of pluripotent stem cells into ectoderm [18]. However, even though laminin-1 is present in all epithelia during development, it is mostly replaced by other laminins as the epithelium matures. One example of this replacement is seen during glomerulogenesis. During development of nephrons, epithelial cells adjacent to the capillaries of the glomerulus differentiate into podocytes that help establish the glomerular filtration barrier [19]. The podocytes lie on a basement membrane, and the endothelial cells produce a separate basement membrane. During maturation of the glomerulus, these two basement membranes fuse to form the glomerular basement membrane (GBM) [20]. Throughout the early stages of GBM maturation, laminin-1 is present, but it is later replaced by laminins-10 and -11. In the final stages of maturation, laminin-10 disappears, leaving laminin-11 as the sole laminin in the GBM [21].

One of the cellular effects of laminin-1 is seen in experiments with cells of the dorsal root ganglia. In the presence of laminin-1 in the matrix, interactions with α3β1 and α7β1 integrins on the cells lead to neurite outgrowths [22]. It thus seems that the signaling induced by laminin-1 is mediated through interaction with other proteins such as cell surface receptors that relay information and ultimately alter gene transcription.

LAMININ-2

The major laminin in skeletal muscle is laminin-2 [23]. The integrity of muscle cells is ensured by the dystrophin–glycoprotein complex that mediates crucial linkages [24]. It is believed that α-dystroglycan binds the G domain of the laminin α2 chain, thus stabilizing the basement membrane and enabling laminin polymerization [13–15,25]. Another important interaction is seen at myotendinous junctions where the laminin α2 chain interacts with α7β1 integrin. Absence of the integrin α7 subunit leads to a form of muscular dystrophy [26,27].

In the neuromuscular system of mammals, seven laminin chains are expressed (α1, α2, α4, α5, β1, β2, and γ1) that produce laminins-1, -2, -4, -8, -9, -10, and -11. In-depth discussion of the distribution of these laminins in the various basal laminae is beyond the scope of this chapter, but it has been reviewed previously [23,28]. In Schwann cells of peripheral nerves, laminin-2 is present in the basal lamina [29]. The interaction of laminin-2 with α-dystroglycan is also seen in axons of the peripheral nervous system where it is important for degeneration and regeneration of axons. The expression of the laminin α2 chain correlates with levels of the cytoplasmic domain of β-dystroglycan [30]. With increasing axonal deterioration, immunoreactivity of both the laminin α2 chain and β-dystroglycan decreased, but progressively increased during regeneration [30]. Based on these findings, it was speculated that the interaction between the dystroglycan complex and laminin-2 enables myelin formation.

Laminin-2 has also been found to be part of the mesangial matrix of the kidney in mice and humans, but not in rats [21,31–33].

LAMININ-3

The myotendinous junctions are specialized sites for the transmission of force between the muscle and tendon. During formation of myotendinous junctions, a special basement membrane is formed where laminins α1 and α5 chains (ie, LAMA1 and LAMA5) are specifically localized. Later during development, the laminin β2 chain (LAMB2) is expressed, and LAMA5 becomes distributed along the entire myotube. Based on the codistribution of the laminin chains, it was speculated that a switch from laminin-1 to laminin-3 occurs during development of the myotendinous junction [34].

LAMININ-4

Although laminin-2 is the major laminin present in skeletal muscle, laminin-4 is also present and resides in the neuromuscular junction [23]. It has been shown that mice lacking LAMB2 have severe functional defects in the neuromuscular junction and die within a few weeks after birth [35]. Through studies it has become clear that this effect is mediated by an apparent repellent effect of LAMB2 on the Schwann cells [36]. In the presence of LAMB2, the Schwann cell acted normally, but in its absence, the Schwann cell wrapped the nerve terminal in membrane extensions that obstructed the synaptic cleft and blocked the passage of the neurotransmitter from nerve to muscle.

Little evidence exists for the presence of laminin-4 in the mesangial matrix of the kidney. However, in comparison to humans and mice, which have laminin-2 in their mesangial matrix, this matrix contains laminin-4 in rats [21,31–33].

LAMININS-5, -6, AND -7

In general, epithelial tissues line all surfaces facing the outer environment and include the skin, oral cavity, lung, hair follicles, many glands, and the gastrointestinal and urinary tracts. All epithelial cells are polarized and their basolateral surface rests on a basement membrane. Epithelial basement membranes contain a high level of laminin-5 that interacts with a range of different proteins, such as the α6β4 integrin, and the hemidesmosome through interaction with BP-180 and intracellular plectin, both of which are part of the hemidesmosome [28]. Attachment of the basement membrane to the underlying stroma is partially achieved through the interaction of laminin-5 with type VII collagen, which forms part of the anchoring fibrils that extend into anchoring plaques in the stroma [37–39]. The epithelial basement membrane zone consists of a range of structures maintaining the stable association of the epidermis and the dermis. Among others, these structures include the aforementioned hemidesmosomes and anchoring fibrils. Combined, these structures form a highly interconnected complex, thereby securing correct organization of the layers of the skin. If one, or more, of the proteins in this meshwork are altered, eg, due to genetic alteration, the skin becomes fragile, often leading to one of the types of epidermolysis bullosa [40].

The epithelial basement membrane also contains laminin-6 and, to a lesser extent, laminin-7. Through studies of the epithelial basement membrane of the amnion, it seems that laminins-6 and -7 form a covalent complex with laminin-5 [41]. Because the basement membrane of the amnion resembles that of the skin, this interaction is likely to be present in the skin. During mechanical stimuli, laminin-6 has been shown to be involved in signal transduction through complexes formed with perlecan and dystroglycan [42].

LAMININ-8

Laminin-8 (α4β1γ1) is expressed by all endothelial cells and is present in the basement membrane of the vasculature, along with other laminins such as -9, -10, and -11. The crucial role of the laminins containing the laminin α4 chain (LAMA4) in the basement membrane of the endothelium, and their role in microvessel formation, was shown through deletion of the chain in mice [43]. Both during the embryonic and neonatal period, *Lama4* null mice presented with diffuse bleeding in subcutaneous fat and also intramuscularly [43]. Mutation of *Lama4* led to delayed deposition of nidogen and collagen IV into the basement membrane of the capillaries, thus causing discontinuities. The mutation, however, only led to a minor increase in neonatal mortality [43]. Laminin-8 has also been implicated in lymphocyte physiology; it has been shown to be synthesized by lymphoid cells. It also promotes lymphocyte migration and is able to stimulate T-cell proliferation in concert with CD3/T cell receptor engagement [44].

Differentiation of adipocytes from a preadipocyte cell line led to an increased expression of mRNA corresponding to laminin-8 and was thus judged to be part of the ECM of adipose tissue [45]. To determine the effect of LAMA4 ablation on weight gain and adipose tissue function, a study was conducted in mice deficient in *Lama4* (*Lama4*$^{-/-}$) and compared with wild type. In the absence of LAMA4, mice had decreased epididymal adipose tissue mass and defective lipogenesis, but no differences were observed in subcutaneous adipose tissue [46]; thus, it was suggested that LAMA4 affects adipose structure and function in a deposit-specific manner.

In the peripheral nervous system, axons are covered with laminins containing the laminin β1 chain (LAMB1) [23]. This covering is provided by laminins-2, -8, and -10. Through knockout studies, it became clear that LAMA4-containing laminins (eg, laminins-8 and -9) are important for Schwann cell maturation. Whereas heterozygous animals had partial demyelination, animals deficient in LAMA4 had an almost complete amyelination and severe neuropathy [47,48].

LAMININ-9

Laminin-9 is present in the neuromuscular junction [23]. As with laminin-4, the absence of LAMB2 leads to severe muscular defects due to the hindrance of neurotransmitter deliverance to muscle [35,36]. Furthermore, just as absence of LAMA4 leads to improper localization of synapses; it seems that LAMB2 is also involved in guiding proper localization of synapses [49].

A study of mesangial cells suggested that the presence of laminin-9 is necessary for migration induced by insulin-like growth factor binding protein-5 [50]. The migration was inhibited by antibodies that blocked the attachment to α6β1 integrins. Antisense inhibition of laminin-9 expression led to absence of the integrin on the cell surface [50]. These findings suggest that the presence of laminin-9 in the mesangium is necessary for development.

As mentioned for laminin-8, laminins containing LAMA4 (ie, laminins-8 and -9) are required for correct microvessel formation [43]. In the absence of LAMA4, mice developed diffuse and extensive bleeding in subcutaneous fat and also intramuscularly [43]. The deposition of nidogen and type VI collagen into the basement membranes was delayed, which suggests a vital role for the LAMA4 chain in correct basement membrane formation in the endothelium.

LAMININ-10

In a study performed by Pouliot et al. [51], the LAMA5 chain of laminins-10 and -11 was shown to be abundantly expressed in the basement membrane underlying the interfollicular epidermis and in the blood vessels in the dermis of humans [51]. Although laminin-5 expression did not change with age in the basement membrane underlying the epidermis, quantities of laminins-10 and -11 both seemed to decrease in adult skin [51]. It was also proven that these laminins were potent adhesive substrates for keratinocytes, with the adhesion being mediated by interactions with integrins $\alpha3\beta1$ and $\alpha6\beta4$, and that the laminins stimulated keratinocyte proliferation and migration [51].

During hair germ elongation, it has been shown that there is a reduced expression of laminins-1 and -5 [52,53]. Because laminin-10 is present in the basement membrane of elongating hair germs, whereas other laminins are downregulated, Li et al. [54] proposed that laminin-10 was necessary for hair follicle development. To prove this hypothesis, they generated LAMA5-deficient mice and showed that their skin contained fewer hair germs than controls and that the skin presented with lower levels of early hair markers such as sonic hedgehog and Gli1. After transferring skin grafts to nude mice, the lack of hair development could be reversed by the addition of purified laminin-10, thus underlining the vital role of this laminin in hair development [54]. In summary, it seems that laminin-10 is necessary for hair follicle epithelial invagination (ie, papillae formation), but is not required for formation of papilla cells, indicating that this laminin is part of a scaffold necessary for the generation of papillae [55].

The pleura of the lungs are made up of mesothelial cells and an underlying basement membrane containing laminin-10. Null mutation of *Lama5* leads to breakdown of the basement membrane with a resulting inability to separate the four lobes of the right lung [56]. Laminin-10 is also present in various areas of vasculature. In embryos lacking LAMA5, only minor vascular defects were observed. However, the absence of LAMA5 in mice led to other severe defects such as exencephaly, syndactyly, and placentopathy of the embryos [57]. In contrast to the *Lama5* mutation, absence of either LAMB1 or the laminin $\gamma1$ chain (LAMC1) causes early death of the embryo because no basement membrane is able to form [17,58].

As mentioned for laminin-1, laminin-10 is present during development of the GBM [21]. However, during maturation of the GBM it is gradually replaced by laminin-11. Laminin-10 is still found extensively in the mature kidney, but it is localized to the basement membranes of the tubules and collecting ducts [21,31].

A potential effect of laminin-10 during development of neurons was seen in an in vitro study of cells from dorsal root ganglia of mice [22]. It was shown that laminin-10 bound integrin $\alpha6\beta1$ and that this lead to outgrowth of neurites, independently of the addition of the trophic factor neurotrophin [22].

LAMININ-11

Laminin-11 is present in the neuromuscular junctions [23]. In the synaptic basal laminae, LAMA5 and LAMB2 are present, making laminin-11 the prominent laminin of synapses [29,32]. In the absence of LAMB2, severe muscular defects are observed as the delivery of the neurotransmitter is obstructed by membrane extensions of Schwann cells [35,36]. The presence of laminin-11, along with other laminins containing LAMB2, at the neuromuscular junctions thus seems to ensure the integrity of muscle innervations.

As stated for laminin-1, during the early phases of glomerulogenesis laminin-1 is present, but it is gradually replaced by laminins-10 and -11. During the final maturation of the glomerulus, laminin-10 levels decrease, leaving laminin-11 as the sole laminin of the GBM [21]. Immunoelectron microscopy of developing mouse kidneys showed that laminins-1 and -11 were produced by both endothelial cells and podocytes, suggesting that these cell types are important for the development of the GBM [59].

LAMININ-12

A novel laminin was isolated via a precipitation experiment of human placental chorionic villi; it consisted of laminin chains α2, β1, and an unknown γ chain [60]. On the basis of sequence analysis results, the laminin α chain was identified as LAMA2, despite the fact that a monoclonal antibody targeting LAMA2 was unable to recognize it. The laminin β chain was identified using a monoclonal antibody specific to LAMB1 [60]. The identification of the laminin γ proved to be more difficult because it contained sequences that were not found in any of the known laminin chains. Thus, Koch et al. concluded that they had found a novel laminin γ chain that they named laminin γ3 (LAMC3) [60]. The unique combination of chains generated laminin-12.

LAMININS-13, -14, AND -15

The expression profiles of laminins in different areas of the nervous system have been investigated in multiple studies. In the CNS, laminins-13 and -14, but not -15, are present at synaptosomes in, eg, the hippocampus [61]. In the inner retina, laminins have been seen in association with retinal ganglion cells [62–65], and LAMB2 was seen in association with photoreceptors [66,67]. In vitro studies revealed that LAMB2 promotes the phenotype of rod photoreceptors [66,68] and that it is vital for the correct development of photoreceptors [69]. Libby et al. [70] replicated the previous histological findings for the LAMB2 distribution and a similar distribution was also found in the human retina [70]. Isolation of laminin complexes from the retina revealed the presence of two novel laminin heterotrimers termed laminins-14 (α4β2γ3) and -15 (α5β2γ3) [70]. Based on the findings, no conclusive data were presented, but it has been speculated that LAMB2 may be localized to synapses in the CNS as it is in the peripheral nervous system [71].

In the remainder of this chapter, tables representing findings for the individual chains are presented.

LAMA1

Laminin Subunit α1	Description	References
Gene name and number	*LAMA1* UniProt: P25391 Chromosome: 18p11.31	Gene ID: 284217
Mutations with diseases in humans	Mutations in this gene may be associated with Poretti–Boltshauser syndrome. Mutations in *LAMA1* cause cerebellar dysplasia and cysts with or without concomitant retinal dystrophy.	[72]
Null mutation in mice	Null mutations are not viable because they cause death at E6.5–E7 due to the absence of the Reichert membrane. However, conditional *Lama1* knockout mice in the epiblast lineage have been produced in Sox2-Cre mice. This led to hindered development of the cerebellum, which might partially be explained by the resulting conformational defects of the pial membrane. A marked reduction in numbers of dendritic processes in Purkinje cells was also observed in these mice. Adult Sox2-Cre–induced conditional *Lama1* knockout mice developed focal glomerulosclerosis and proteinuria with age. Another model revealed that *Lama1* knockout in mice led to a reduced integrity of the inner limiting membrane of the vitreous, because the vitreal astrocytes migrated into the vitreous.	[17,73–75]
Tissue distribution in healthy states	Present in developing epithelia, myotendinous junctions, and developing lung. It is also important for development of the cerebellum of mice, but does not generally seem to be associated with this tissue.	[73]
Tissue distribution in pathologically affected states	Missense mutation in *Lama1* of mice (Y265C) led to a disruption of retinal vascular development and a thinning of the inner limiting membrane.	[75]
	Enhanced and differential distribution of LAMA1 and LAMC1 distinctly associate these two laminins with Alzheimer disease.	[76]
Special domains	The laminin α chains are different from all other laminin chains because they contain a G domain. The G domain consists of five self-folding modules called laminin G modules (LG1–5). All α chains also contain a long coiled-coil domain (I/II) that mediates heterotrimer assembly with similar domains in the β and γ chains. The laminin α1, α2, α3B, and α5 chains each contain a short arm that consists of alternating globular domains (IVa, IVb, and VI/laminin N-terminal) and rod-like domains composed of laminin EGF-like repeats (IIIa, IIIb, and V) (see Fig. 29.1).	[7]

Laminin Subunit α1	Description	References
Special neoepitopes	None found in literature.	
Protein structure and function	See special domains.	
Binding proteins	Dystroglycan (α chain), heparin, and sulfatides. LAMA1 forms heterotrimers with laminin β and γ chains (see Table 29.1).	[13,15,77,78]
Known central function	Present during glomerulogenesis, but is not part of the mature kidney.	[79]
Animal models with protein affected	See "Null mutation in mice."	[73–75]
	A mouse model expressing an altered LAMA1 version, lacking the globular domains 4 and 5, developed as a fetus, but died due to failed epiblast polarization.	[80]
	In mice with collagen α3 (IV) chain knockout, resulting in Alport syndrome, an upregulation of LAMA1 was seen in the GBM. The increased levels of LAMA1 and LAMA5 in the GBM contributed to an abnormal permeability.	[81]
Biomarkers	None found in the literature. However, if the upregulation of LAMA1 in the GBM of the mouse model for Alport syndrome holds true for humans, LAMA1 could be a potential biomarker of Alport syndrome.	[81]

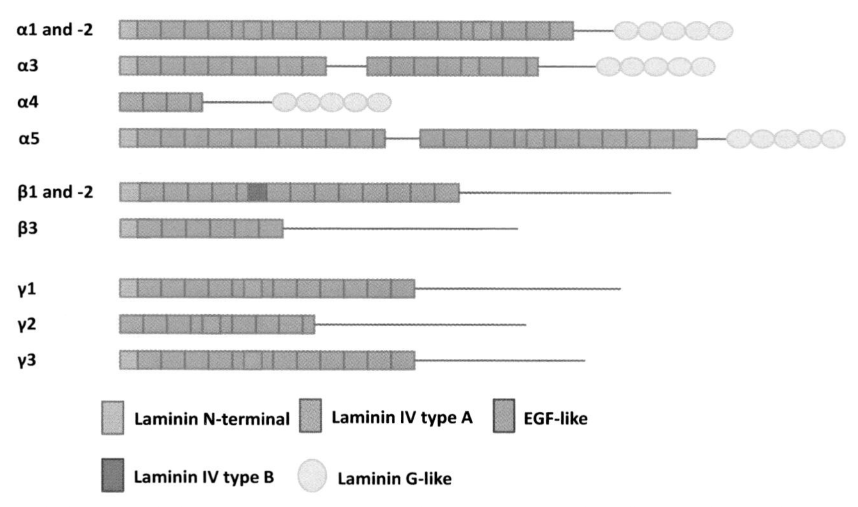

FIGURE 29.1 **Schematic representation of domains in the laminin chains.** Different isoforms of the chains exist. For example, in the representation LAMA3 is presented as the LAMA3B version. In contrast, the LAMA3A lacks the N-terminal 1620 residues and has an altered sequence from residues 1621 to 1665.

LAMA2

Laminin Subunit α2	Description	References
Gene name and number	*LAMA2* UniProt: P24043 Chromosome: 6q22-23	Gene ID: 3908
Mutations with diseases in humans	Associated with merosin-deficient congenital muscular dystrophy and limb-girdle muscular dystrophy. Dilated cardiomyopathy with conduction defects have been noted in a patient with partial merosin (laminin-2) deficiency. In an observational cohort study, *LAMA2* mutations were associated with cardiac problems.	[82–85]
Null mutation in mice	Null mutation in mice leads to growth retardation and severe muscular dystrophic symptoms and mice die by 5 weeks of age. Null mutations in *Lama2* expressed during lung development have not been reported to cause lung defects. The models for total *Lama2* knockout all seem to result in death within the first 15 weeks after birth. All models targeting *Lama2* have varying degrees of muscular dystrophy and neuropathy. There also seems to be defective spermatogenesis and odontoblast differentiation.	[86–89]
Tissue distribution in healthy states	LAMA2 is found in laminins-2, -4, and -12. Laminin-2 is a constituent of all myogenic tissues and peripheral nerves. Laminin-4 is part of the neuromuscular junction and mesangial matrix of glomerulus. Laminin-12 is found in ciliated epithelia, but the existence of this laminin is controversial.	[30]
	Based on levels of mRNA in the myocardium, there seems to be a negative correlation between levels of LAMA4 and age.	[90]
Tissue distribution in pathologically affected states	There seems to be an imbalance between myocyte hypertrophy and the level of LAMA2. The imbalance might contribute to alterations in sarcolemmal properties that occur during the development of cardiac hypertrophy and its transition to cardiac failure.	[90]
Special domains	The laminin α chains are different from all other laminin chains because they contain a G domain. The G domain consists of five self-folding modules called laminin G modules (LG1–5). All α chains also contain a long coiled-coil domain that mediates heterotrimer assembly. Similar domains are found in the β and γ chains. The laminin α1, α2, α3B, and α5 chains each contain a short arm that consists of alternating globular domains and rod-like domains composed of laminin EGF-like repeats (see Fig. 29.1).	[7]
Special neoepitopes	None found in the literature.	

Laminin Subunit α2	Description	References
Protein structure and function	Not determined. The domains of the chain are shown in Fig. 29.1.	
Binding proteins	Integrin α7β1, dystroglycan (α chain), and heparin. LAMA2 forms heterotrimers with laminin β and γ chains (see Table 29.1).	[14]
Known central function	LAMA2 binds to the dystrophin–glycoprotein complex that is the mediator of the linkage between the sarcolemma to both the basement membrane and to the underlying cytoskeleton	[14]
Animals models with protein affected	Hypomorphic mutation of *Lama2* in mice has produced a protein that is missing a part of the N-terminal domain VI. This domain is necessary for trimer formation and the laminin is thus not able to polymerize.	[26,91]
Biomarkers	None found in the literature.	

LAMA3

Laminin Subunit α3	Description	References
Gene name and number	LAMA3 UniProt: Q16787 Chromosome: 18q11.2	Gene ID: 3909
Mutations with diseases in humans	Mutations in the protein have been shown to cause epidermolysis bullosa.	
	Reverse transcription-polymerase chain reaction analysis of mRNA extracted from the proband's keratinocytes identified a homozygous single base pair deletion in the transcripts encoding the laminins α3A and 3B isoforms.	[92]
	Nonsense mutations within *LAMA3* have been shown to be associated with the pathogenesis of junctional epidermolysis bullosa.	[93]
Null mutation in mice	Targeted disruption of *Lama3* in mice reveals abnormalities in survival and late-stage differentiation of epithelial cells. Severe skin blistering and abnormal glomerulogenesis were noted.	[94,95]
Tissue distribution in healthy states	LAMA3 is found in laminins-5, -6, and -7 of the epidermis.	
	LAMB3 is also part of laminin-13 found in the CNS at synapses in the hippocampus.	[61]
	In the lungs, LAMA3 is expressed by alveolar epithelial cells.	[96]
Tissue distribution in pathologically affected states	Disruption of *LAMA3* leads to disorganized hemidesmosomes. The malfunction in crosslinking at the basement membrane between the epidermis and the dermis leads to fragility and blister formation of the skin.	[40]
	Production of laminin-5 (α3β3γ2) is uncoordinated in the airway epithelium of allergic asthmatics.	[97]

Laminin Subunit α3	Description	References
Special domains	The laminin α chains are different from all other laminin chains because they contain a G domain. The G domain consists of five self-folding modules called laminin G modules (LG1–5). All α chains also contain a long coiled-coil domain that mediates heterotrimer assembly. Similar domains are found in the β and γ chains. The laminin α1, α2, α3B, and α5 chains each contain a short arm that consists of alternating globular domains and rod-like domains composed of laminin EGF-like repeats (see Fig. 29.1). Two well-described LAMA3 chains are presented in the literature: the A-chain and B-chain. The A-chain differs from the B-chain by the absence of the N-terminal 1620 residues and an altered sequence from residue 1621 to 1665.	[7]
Special neoepitopes	None found in the literature.	
Protein structure and function	The laminin α3 chain resembles a truncated version of the laminin α1 and α2 chains. To see a structural representation of the differences between these and other chains, see Fig. 29.1.	[7]
Binding proteins	α1β3 integrin, plasminogen, and tissue plasminogen activator exhibit specific high affinity binding to the LG1 subdomain of the α3 chain of laminin. Other binding proteins are syndecan-2 and -4 and heparin. LAMA3 forms heterotrimers with laminin β and γ chains (see Table 29.1).	[98–100]
Known central function	Part of the complex that forms hemidesmosomes of the epidermis and thus keeps this structure stable.	[40]
Animal models with protein affected	Targeted disruption has been shown to lead to profound epithelial abnormalities, causing neonatal death of the affected mice.	[94]
Biomarkers	No biomarker as such has been developed, but LAMA3 could be of potential interest as high LAMA3 expression is associated with gastric cancer.	[101]

LAMA4

Laminin Subunit α4	Description	References
Gene name and number	LAMA4 UniProt: Q16363 Chromosome: 6q21	Gene ID: 3910
Mutations with diseases in humans	Mutations of LAMA4 have been shown to give rise to cardiomyopathy, a disorder characterized by ventricular dilation and impaired systolic function. The mutations ultimately result in congestive heart failure and arrhythmia with a concomitant increase in risk of premature death. The disease is caused by defects in cardiomyocytes and endothelial cells.	[102]

Laminin Subunit α4	Description	References
Null mutation in mice	During the embryonic and neonatal period, *Lama4* knockout mice presented with hemorrhaging. Extensive bleeding, decreased microvessel growth and mild locomotion defects were also seen in experimental settings to monitor angiogenesis. Mice deficient in LAMA4 had an almost complete amyelination of nerves and presented with severe neuropathy.	[43,47–49]
Tissue distribution in healthy states	In healthy tissue, the LAMA4 chain is present in endothelium, smooth muscle (laminins-8 and -9), adipose tissue, peripheral nerves (laminin-8), neuromuscular junctions (laminin-9), mesangial matrix of the glomerulus (laminin-9), CNS and retina (laminin-14) (see Table 29.1). It is also found in the lungs where the laminin α4 chain is expressed by lung fibroblasts.	[70,96]
Tissue distribution in pathologically affected states	LAMA4 is highly expressed in human hepatocellular carcinoma in Chinese patients and is thought to be a novel marker of tumor invasion and metastasis. LAMA4 is also widely expressed in renal cell carcinomas and has a deadhesive effect. Downregulation of LAMA4 expression inhibits glioma invasion in vitro and in vivo. The epithelial-to-mesenchymal transition has been shown to downregulate LAMA5 and upregulate LAMA4 in oral squamous carcinoma cells, which could promote cancer cell invasion.	[96,103,104]
Special domains	The laminin α chains are different from all other laminin chains as they contain a G domain. The G domain consists of five self-folding modules called laminin G modules (LG1–5). All α chains also contain a long coiled-coil domain that mediates heterotrimer assembly. Similar domains are found in the β and γ chains. To see a representation of the domains found in LAMA4, see Fig. 29.1.	[7]
Special neoepitopes	The high expression of LAMA4 in different cancers could potentially be used to find neoepitopes that are more abundantly present in these pathological settings (see "Biomarkers").	
Protein structure and function	The overall size of the LAMA4 chain is much smaller than that of other laminin α chains. It has four EGF-like domains, but the fourth is truncated. It still contains five LG domains (see Fig. 29.1).	
Binding proteins	LAMA4 interacts with αVβ3 integrin, heparin, syndecans-2 and -4.	[105,106]

Laminin Subunit α4	Description	References
Known central function	LAMA4 is thought to be implicated in the development of embryos, but the exact function is not known. The involvement of LAMA4 in development could be through its ability to promote microvessel formation.	[105,106]
	LAMA4 has been shown to bind to syndecans-2 and -4 and promote cell adhesion.	[43]
	LAMA4 binds heparin through the LG4 domain.	[106]
	LAMA4 is also thought to stimulate angiogenesis by its interaction with the αVβ3 integrin, which is mediated by residues 1121–1139 of the LG2 domain.	[105,106]
	Deficiency of Lama4 showed that this chain is involved in lipogenesis of epididymal adipose tissue, but not in lipogenesis of subcutaneous adipose tissue.	[46]
Animal models with protein affected	Null mutation in mice resulted in impaired microvessel growth.	[43]
	Nude mice with an intracerebral injection of glioma cells transfected with an antisense oligonucleotide for *LAMA4* resulted in the formation of a noninvasive tumor, whereas injection of glioma cells transfected with sense oligonucleotide for *LAMA4* resulted in diffuse invasion of brain tissue. It thus seems that expression of LAMA4 promotes invasion of glioma cells both in vitro and in vivo.	[107]
Biomarkers	LAMA4 is a potential biomarker for invasion and metastasis in hepatocellular carcinoma.	[108]
	As LAMA4 has also been seen in association with cells from oral squamous carcinoma and renal cell carcinoma, it might be interesting to assess this chain for its potential as a diagnostic tool for various types of cancer.	[103,104]

LAMA5

Laminin Subunit α5	Description	References
Gene name and number	LAMA5 UniProt: O15230 Chromosome: 20q13.2–13.3	Gene ID: 3911
Mutations with diseases in humans	No mutations have been described in the literature.	
Null mutation in mice	Elimination of *Lama5* results in multiple defects, eg, in glomerular filtration and late gestational lethality due to placental defects. The LAMA5 deficiency also demonstrated a role for laminins containing this chain during development of the GBM, septations in the lung, hair follicles, dental epithelium, and submandibular gland morphogenesis.	[54,56,57, 109–111]

Laminin Subunit α5	Description	References
Tissue distribution in healthy states	LAMA5 is found in mature epithelium, mature endothelium, smooth muscle, neuromuscular junction, GBM, and the CNS and retina. During glomerulogenesis, there is a transition in laminin gene expression from laminin-1 and laminin-10 to laminin-11 as the GBM is maturing. In the GBM, laminin-11 forms an intricate meshwork with collagen type IV, perlecan, and nidogen.	[70,109]
	In the lungs, LAMA5 is expressed by alveolar epithelial cells.	[96]
Tissue distribution in pathologically affected states	In collagen α3 (IV) knockout mice—the Alport model—laminin α5 was highly present in the thickened GBM and led to an abnormal permeability. It seems that the LAMA5 LG1–2 tandem domain plays an essential role during development and harbors the majority of the functionality of the LAMA5 LG domain. Also, the LAMA5 LG3–5 tandem serves as a novel determinant required for the kidney's glomerular filtration barrier to plasma protein.	[81,112]
Special domains	The laminin α chains are different from all other laminin chains because they contain a G domain. The G domain consists of five self-folding modules called laminin G modules (LG1–5). All α chains also contain a long coiled-coil domain that mediates heterotrimer assembly. Similar domains are found in the β and γ chains. The laminin α1, α2, α3B, and α5 chains each contain a short arm that consists of alternating globular domains and rod-like domains composed of laminin EGF-like repeats (see Fig. 29.1).	[7]
	LAMA5 is the largest laminin chain (see Fig. 29.1). Although it contains the same types of domains as the other laminin α chains, it has been shown that some of the specific LAMA5 domains are essential for development. Through a series of knock-in studies in *Lama5*-deficient embryos, it was shown that LG1–2 were vital for development and that the LG3–5 domains were necessary for correct assembly of the glomerular filtration barrier of the kidney.	[112]
Special neoepitopes	LAMA5 is cleaved by MT1-MMP.	[113]
Protein structure and function	Thought to be part of basement membranes in the tissues mentioned above and functions as a stabilizing agent.	
Binding proteins	Integrins α3β1, α6β1, α6β4, and Lutheran glycoprotein bind to LAMA5.	[12,114]
	LAMA5 forms heterotrimers with laminin β and γ chains (see Table 29.1).	

Laminin Subunit α5	Description	References
Known central function	Part of the GBM where it is thought to stabilize and regulate permeability.	[115]
Animal models with protein affected	Due to the lethality of total deficiency of laminin α5 chains, a chimeric animal model was created with the traditional LAMA5 G domain being swapped for that of LAMA1. Through various knockin experiments, it was shown that the LG1–2 domains are vital for development and that the LG3–5 domains are necessary for correct assembly of the glomerular filtration barrier of the kidney.	[112]
	A podocyte-specific LAMA5 deficiency was created by Cre-lox and tetO. This led to varying degrees of proteinuria and rates of progression to nephrotic syndrome. Restoration of podocyte-specific LAMA5 expression led to restoration of normal function. The specific ablation of LAMA5 in podocytes with concomitant development of proteinuria suggests that LAMA5 expression by podocytes is necessary for maintenance of GBM integrity.	[116]
	A study in mice showed that a hypomorphic mutation caused polycystic kidney disease.	[117]
	In a mouse model of Alport syndrome, knockout of collagen α3 (IV) led to upregulation of LAMA5 in the GBM.	[81]
Biomarkers	If the upregulation of LAMA5 in the mouse model for Alport syndrome is also true for humans, LAMA5 could be a potential biomarker of Alport syndrome.	[81]
	Also, because MT1-MMP is able to cleave LAMA5, and prostate cancer cells have been shown to express MT1-MMP, fragments of LAMA5 could be potential biomarkers in prostate cancer.	[113]

LAMB1

Laminin Subunit β1	Description	References
Gene name and number	LAMB1 UniProt: P07942 Chromosome: 7q22	Gene ID: 3912
Mutations with diseases in humans	Mutations in *LAMB1* cause cobblestone brain malformation without muscular or ocular abnormalities. The disease is an autosomal recessive brain malformation characterized by cobblestone changes in the cortex that are more severe in the posterior region, and subcortical band heterotopias. Affected individuals have hydrocephalus, seizures, and severely delayed psychomotor development.	[118]

Laminin Subunit β1	Description	References
Null mutation in mice	Knockout of this gene results in early death of mice at around E5.5. The lethality is due to absence of basement membrane formation.	[17]
Tissue distribution in healthy states	Developing and mature epithelium (laminins-1, -6, and -10), myogenic tissue, peripheral nerves (laminins-2 and -8) endothelium (laminins-8 and -10), smooth muscle (laminin-10), and adipose tissue.	
Tissue distribution in pathologically affected states	It has been demonstrated that LAMB1 and integrin α2 expressions are elevated in the anterior temporal neocortex tissue from patients with intractable epilepsy.	[119]
Special domains	The LAMB1 and LAMB2 chains share similar domains. The LAMB3 chain, however, resembles a truncated version and does not contain a laminin IV type B domain (see Fig. 29.1).	
Special neoepitopes	None found in the literature.	
Protein structure and function	The three-dimensional structure of LAMB1 has not been identified.	
	As null mutation of *Lamb1* in mice leads to early death. The LAMB1 chain plays an important role during early development.	[17]
Binding proteins	LAMB1 forms heterotrimers with laminin α and γ chains, thereby forming laminins-1, -2, -6, -8, -10, and -12 (see Table 29.1). LAMB1 has also been predicted to interact with collagen type XVII and integrins α1, α6, and α7.	
Known central function	LAMB1 is an integral part of most basement membranes of the body. Other functions: See above.	
Animal models with protein affected	None found in the literature.	
Biomarkers	None found in the literature.	

LAMB2

Laminin Subunit β2	Description	References
Gene name and number	LAMB2 UniProt: P55268 Chromosome: 3p21	Gene ID: 3913
Mutations with diseases in humans	LAMB2 deficiency causes Pierson syndrome, a congenital nephrosis with mesangial sclerosis and distinct eye abnormalities. Hypomorphic mutations in *LAMB2* cause milder phenotypes of Pierson syndrome.	[120,121]

Laminin Subunit β2	Description	References
Null mutation in mice	A *Lamb2* knockout does not affect the development of the GBM because LAMB1 seems to be able to partially replace LAMB2. However, there are serious functional consequences of the LAMB2 deficiency that lead to defects in the neuromuscular junctions, severe proteinuria, and retinal pathology. It thus seems that LAMB2 is indispensable for the integrity of the GBM.	[35,69,122]
Tissue distribution in healthy states	Myotendinous junctions (laminin-3), neuromuscular junction (laminins-4, -9, and -11), epidermis (laminin-7), endothelium (laminins-9 and -11), smooth muscle (laminins-9 and -11), mesangial matrix of the glomerulus (laminins-4 and -9), GBM (laminin-11), the CNS and retina (laminins-13, -14, and -15).	
Tissue distribution in pathologically affected states	Overexpression of the LAMB2 chain is associated with increased basement membrane thickness and is possibly related to spermatogenic dysfunction.	[123]
Special domains	The LAMB1 and LAMB2 chains share similar domains. In comparison, the LAMB3 chain resembles a truncated version and does not contain a laminin IV type B domain (see Fig. 29.1).	
Special neoepitopes	None found in the literature.	
Protein structure and function	See "Special domains."	
Binding proteins	α3β1 and α7X2β1 integrins bind more avidly to LAMB2 containing laminins than LAMB1 containing laminins. LAMB2 also interacts with nidogen. LAMB2 forms heterotrimers with laminin α and γ chains (see Table 29.1).	[124,125]
Known central function	Components of laminin heterotrimers in basement membrane (see "Tissue distribution in healthy states").	
Animal models with protein affected	LAMB2 deficiency causes the glomerular filtration barrier to fail, leading to severe proteinuria.	[122]
	Mutations of *Lamb2* have shown that it regulates formation of motor nerve terminals.	[35]
Biomarkers	None found in the literature.	

LAMB3

Laminin Subunit β3	**Description**	**References**
Gene name and number	LAMB3 UniProt: Q13751 Chromosome: 1q32	Gene ID: 3914
Mutations with diseases in humans	Epidermolysis bullosa, junctional epidermolysis bullosa, Herlitz-type generalized atrophic benign epidermolysis bullosa	[126–129]
Null mutation in mice	Not viable as it causes perinatal death. Deficiency was achieved through spontaneous insertion of an intracisternal-A particle in the gene encoding LAMB3. Homozygous mice survived for less than 30 h after birth and had very fragile skin, which resulted in dislodgement of the entire epidermal layer of the paws, limbs, tail tips as well as causing blisters of the skin. The model resembles junctional epidermolysis bullosa.	[130,131]
Tissue distribution in healthy states	Found at the dermal-epidermal junction	[132]
Tissue distribution in pathologically affected states	Laminin-5 ($\alpha3\beta3\gamma2$) staining is attenuated in patients with the mutations that cause junctional epidermolysis bullosa. Uncoordinated production of laminin-5 occurs in lungs of allergic asthma patients, which might play a role in the poor epithelial cell anchorage in these patients.	[97,133]
	In biliary cancer LAMB3; LAMC2, which constitutes two of three chains of laminin-5; and MMP-7 are overexpressed.	[134]
Special domains	None found in the literature. LAMB3 is the smallest of the laminin β chains (see Fig. 29.1).	
Special Neoepitopes	LAMB3 is cleaved by MT1-MMP.	[135]
Protein structure and function	Cell surface collagen XVII can interact with laminin-5 and together they participate in the adherence of a cell to the ECM.	[129]
Binding proteins	ECM1 interacts with fibulin-3 and the LAMB3 chain of laminin-2. LAMB3 forms heterotrimers with laminin α and γ chains (see Table 29.1).	[136]
Known central function	See "Protein structure and function."	
Animal models with protein affected	A mouse model for junctional epidermolysis bullosa has been developed.	[137]

Laminin Subunit β3	Description	References
Biomarkers	Laminin-5 (α3β3γ2) is a biomarker of invasiveness in cervical adenocarcinoma.	[138]
	Because LAMB3 has been shown to be a prometastatic gene during lung cancer, it could have potential as a biomarker to detect metastatic lung cancer.	[134]
	MT1-MMP has been shown to be upregulated in prostate cancer. Because it is also able to cleave LAMB3 in vitro and thereby cause cell migration, it could be interesting to assess the potential of this laminin chain as a biomarker.	[135]

LAMC1

Laminin Subunit γ1	Description	References
Gene name and number	LAMC1 UniProt: P11047 Chromosome: 1q31	Gene ID: 3915
Mutations with diseases in humans	A polymorphism in the promoter of *LAMC1* may increase susceptibility to early-onset pelvic organ prolapse	[139]
Null mutation in mice	Knockout results in embryonic lethality due to failure of endoderm differentiation.	[58]
Tissue distribution in healthy states	LAMC1 is present in most basement membranes (ie, in laminins-1, -2, -3, -4, -6, -7, -8, -9, -10, -11) (see Table 29.1).	
Tissue distribution in pathologically affected states	LAMC1 expression seems to be associated with meningioma grades and could play a role in enhancing tumor invasion.	[140]
	In samples of serous tubal intraepithelial carcinoma an altered staining pattern and increased immunoreactivity of LAMC1 were seen.	[141]
	Enhanced and differential distribution of LAMA1 and LAMC1 distinctly associate these two laminins with Alzheimer disease.	[142]
Special domains	The neurite outgrowth domain of LAMC1 accumulates in the floor plate region of the notochord and in glial fibrillary acidic protein–immunoreactive glial fibers of the embryonic spinal cord.	[143]
Special neoepitopes	None found in the literature.	
Protein structure and function	The nanostructure of the LAMC1 short arm has been determined. In solution, it adapts an extended conformation.	[144]
Binding proteins	Nidogens-1 and -2	[145,146]
	LAMC1 forms heterotrimers with laminin α and β chains (see Table 29.1).	

Laminin Subunit γ1	Description	References
Known central function	LAMC1 is found in most basement membranes. The interaction of LAMC1 with nidogen seems to be necessary for the development of lung and kidney.	[145]
Animal models with protein affected	Specific ablation of the nidogen binding site in the laminin γ1 chain interferes with kidney and lung development. The deletion of the nidogen binding site also leads to cortical dysplasia. Cell type–specific deficiency of LAMC1 in Schwann cells lead to peripheral nerve defects.	[145,147]
Biomarkers	Because of the association with meningioma grade, recurrence and progression-free survival LAMC1 might be a biomarker for this malignancy.	[140]
	Because alterations in staining patterns and an increased immunoreactivity of LAMC1 are seen in serous tubal intraepithelial carcinoma, it could be an interesting target for biomarker development.	[141]
	Because LAMC1, together with LAMA1, are distinctly associated with Alzheimer disease, they might prove to be biomarkers.	[142]

LAMC2

Laminin Subunit γ2	Description	References
Gene name and number	Laminin γ2 chain (LAMC2) UniProt: Q13753 Chromosome: 1q25-31	Gene ID: 3918
Mutations with diseases in humans	Mutations in all of the chains found in laminin-5 (α3β3γ2) can lead to epidermolysis bullosa, junctional epidermolysis bullosa, or Herlitz-type epidermolysis bullosa.	[40]
Null mutation in mice	Null mutations cause perinatal death. Targeted inactivation of *Lamc2* produces a model similar to human junctional epidermolysis bullosa and is characterized by blister formation of the skin.	[148]
	Abnormal tracheal hemidesmosomes were also observed in *Lamc2* null mice.	[149]
Tissue distribution in healthy states	In the epidermis, it is mainly found in the basement membrane within laminin-5 (see Table 29.1). In normal esophageal tissues, LAMC2 is expressed in the basement membrane.	[150]
Tissue distribution in pathologically affected states	Laminin-5, containing LAMC2, has been found to be associated with invasiveness of cervical adenocarcinomas.	[138]
	LAMC2 has also been shown to be increased in the circulation of patients with early-phase acute lung injury.	[151]

Laminin Subunit γ2	Description	References
	LAMC2 may play a role in the progression of esophageal squamous cell carcinoma, and simultaneous expression with secreted protein acidic and rich in cysteine (SPARC) is correlated with poorer prognosis.	[150]
	Uncoordinated expression of laminin-5 is found in the airway epithelium of patients with allergic asthma.	[97]
Special domains	None found in the literature.	
Special neoepitopes	LAMC2 has been shown to be cleaved by MMP-2 between residues 586 and 587, exposing a putative cryptic promigratory site that triggers cell motility.	[152]
	LAMC2 is cleaved by MT1-MMP, causing cell migration.	[153]
	Bone morphogenetic protein-1 can cleave LAMC2.	[154]
Protein structure and function	Part of the laminin heterotrimer (see Table 29.1). Part of the basement membrane of epidermis and esophageal tissues (see Laminin-5 section).	
Binding proteins	Syndecan-1	[155]
	LAMC2 forms heterotrimers with laminin α and β chains (see Table 29.1).	
Known central function	Part of the basement membrane.	
Animal models with protein affected	Targeted inactivation and a hypomorphic mutation of *Lamc2* produce models resembling human junctional epidermolysis bullosa.	[148,156]
Biomarkers	Laminin-5 is a biomarker of invasiveness in cervical adenocarcinoma. Because LAMC2 is a part of laminin-5, it could be a potential biomarker for this type of cancer. LAMC2 is also a promising biomarker for pancreatic cancer and bladder cancer. LAMC2 is a biomarker of acute lung injury.	[138,151, 157,158]

LAMC3

Laminin Subunit γ3	Description	References
Gene name and number	LAMC3 UniProt: Q9Y6N6 Chromosome: 9q34.12	Gene ID: 10319
Mutations with diseases in humans	Recessive *LAMC3* mutations cause malformations during the development of the occipital cortex	[159]
Null mutation in mice	*LAMC3* null mice have no apparent phenotype.	[160]

Laminin Subunit γ3	Description	References
Tissue distribution in healthy states	In situ hybridization of the midfetal human brain results in robust LAMC3 expression in the cortical plate and within the germinal zones of the neocortical wall, including the ventricular zone, subventricular zone, the outer subventricular zone, and the intermediate zone. In contrast, LAMC3 expression in the developing mouse brain is limited to the vasculature and meninges.	[159]
Tissue distribution in pathologically affected states	A homozygous two-base pair deletion was identified *LAMC3*, which led to a premature termination codon. The premature termination resulted in malformation of occipital cortical development. It is not known how this affects the distribution of LAMC3 in pathologically affected states.	[159]
Special domains	Laminin isoforms containing LAMC3 are unable to bind to integrins due to the absence of the glutamic acid residue conserved in the C-terminal regions of LAMC1 and LAMC2.	[161]
Special neoepitopes	None found in the literature.	
Protein structure and function	Part of laminins-12, -13, -14, and -15, which are found in surface of human chorionic placenta villi and basement membrane of the CNS and the retina (see Table 29.1).	
Binding proteins	Nidogens-1 and -2.	[162]
	In contrast to LAMC1 and LAMC2, LAMC3 is not able to bind integrins because of absence of a glutamic acid residue required for this interaction. LAMC3 forms heterotrimers with laminin α and β chains (see Table 29.1).	[161]
Known central function	Has only been found in laminins of the CNS and surface of ciliated cells. It is thus believed that LAMC3 may play an important role in the formation and integrity of the basement membrane of the CNS.	
Animals models with protein affected	Mice homozygous null for *Lamb2* and *Lamc3* exhibit cortical laminar disorganization and exhibit hallmarks of human cobblestone lissencephaly (type II, nonclassical).	[163]
Biomarkers	None found in the literature.	

E, embryonic day; *EGF*, epidermal growth factor; *G domain, large 100-kDa*, C-terminal globular domain; *GBM*, glomerular basement membrane; *LG*, laminin G; *MT1-MMP*, membrane type 1 matrix metalloprotease; *MMP*, matrix metalloprotease; *ECM*, extracellular matrix.

REFERENCES

[1] Timpl R. Structure and biological activity of basement membrane proteins. Eur J Biochem 1989;180:487–502.

[2] Timpl R. Macromolecular organization of basement membranes. Curr Opin Cell Biol 1996;8:618–24.

[3] Poschl E, Schlotzer-Schrehardt U, Brachvogel B, Saito K, Ninomiya Y, Mayer U. Collagen IV is essential for basement membrane stability but dispensable for initiation of its assembly during early development. Development 2004;131:1619–28.
[4] Engel J, Odermatt E, Engel A, Madri JA, Furthmayr H, Rohde H, et al. Shapes, domain organizations and flexibility of laminin and fibronectin, two multifunctional proteins of the extracellular matrix. J Mol Biol 1981;150:97–120.
[5] Macdonald PR, Lustig A, Steinmetz MO, Kammerer RA. Laminin chain assembly is regulated by specific coiled-coil interactions. J Struct Biol 2010;170:398–405.
[6] Beck K, Hunter I, Engel J. Structure and function of laminin: anatomy of a multidomain glycoprotein. FASEB J 1990;4:148–60.
[7] Timpl R, Tisi D, Talts JF, Andac Z, Sasaki T, Hohenester E. Structure and function of laminin LG modules. Matrix Biol 2000;19:309–17.
[8] Tunggal P, Smyth N, Paulsson M, Ott MC. Laminins: structure and genetic regulation. Microsc Res Tech 2000;51:214–27.
[9] Colognato H, Winkelmann DA, Yurchenco PD. Laminin polymerization induces a receptor-cytoskeleton network. J Cell Biol 1999;145:619–31.
[10] Sasaki T, Fassler R, Hohenester E. Laminin: the crux of basement membrane assembly. J Cell Biol 2004;164:959–63.
[11] Hallmann R, Horn N, Selg M, Wendler O, Pausch F, Sorokin LM. Expression and function of laminins in the embryonic and mature vasculature. Physiol Rev 2005;85:979–1000.
[12] Kikkawa Y, Sanzen N, Fujiwara H, Sonnenberg A, Sekiguchi K. Integrin binding specificity of laminin-10/11: laminin-10/11 are recognized by alpha 3 beta 1, alpha 6 beta 1 and alpha 6 beta 4 integrins. J Cell Sci 2000;113(Pt 5):869–76.
[13] Henry MD, Campbell KP. Dystroglycan inside and out. Curr Opin Cell Biol 1999;11:602–7.
[14] Wizemann H, Garbe JH, Friedrich MV, Timpl R, Sasaki T, Hohenester E. Distinct requirements for heparin and alpha-dystroglycan binding revealed by structure-based mutagenesis of the laminin alpha2 LG4-LG5 domain pair. J Mol Biol 2003;332:635–42.
[15] Winder SJ. The complexities of dystroglycan. Trends Biochem Sci 2001;26:118–24.
[16] Aumailley M, Bruckner-Tuderman L, Carter WG, Deutzmann R, Edgar D, Ekblom P, et al. A simplified laminin nomenclature. Matrix Biol 2005;24:326–32.
[17] Miner JH, Li C, Mudd JL, Go G, Sutherland AE. Compositional and structural requirements for laminin and basement membranes during mouse embryo implantation and gastrulation. Development 2004;131:2247–56.
[18] Li L, Arman E, Ekblom P, Edgar D, Murray P, Lonai P. Distinct G. Development 2004;131:5277–86.
[19] Pavenstadt H, Kriz W, Kretzler M. Cell biology of the glomerular podocyte. Physiol Rev 2003;83:253–307.
[20] Abrahamson DR. Origin of the glomerular basement membrane visualized after in vivo labeling of laminin in newborn rat kidneys. J Cell Biol 1985;100:1988–2000.
[21] Miner JH, Patton BL, Lentz SI, Gilbert DJ, Snider WD, Jenkins NA, et al. The laminin alpha chains: expression, developmental transitions, and chromosomal locations of alpha1-5, identification of heterotrimeric laminins 8-11, and cloning of a novel alpha3 isoform. J Cell Biol 1997;137:685–701.
[22] Plantman S, Patarroyo M, Fried K, Domogatskaya A, Tryggvason K, Hammarberg H, et al. Integrin-laminin interactions controlling neurite outgrowth from adult DRG neurons in vitro. Mol Cell Neurosci 2008;39:50–62.
[23] Patton BL. Laminins of the neuromuscular system. Microsc Res Tech 2000;51:247–61.
[24] Cohn RD, Campbell KP. Molecular basis of muscular dystrophies. Muscle Nerve 2000;23:1456–71.

[25] Colognato H, Yurchenco PD. The laminin alpha2 expressed by dystrophic dy(2J) mice is defective in its ability to form polymers. Curr Biol 1999;9:1327–30.
[26] Yao CC, Ziober BL, Squillace RM, Kramer RH. Alpha7 integrin mediates cell adhesion and migration on specific laminin isoforms. J Biol Chem 1996;271:25598–603.
[27] Mayer U, Saher G, Fassler R, Bornemann A, Echtermeyer F, von der MH, et al. Absence of integrin alpha 7 causes a novel form of muscular dystrophy. Nat Genet 1997;17:318–23.
[28] Domogatskaya A, Rodin S, Tryggvason K. Functional diversity of laminins. Annu Rev Cell Dev Biol 2012;28:523–53.
[29] Patton BL, Miner JH, Chiu AY, Sanes JR. Distribution and function of laminins in the neuromuscular system of developing, adult, and mutant mice. J Cell Biol 1997;139:1507–21.
[30] Masaki T, Matsumura K, Saito F, Yamada H, Higuchi S, Kamakura K, et al. Association of dystroglycan and laminin-2 coexpression with myelinogenesis in peripheral nerves. Med Electron Microsc 2003;36:221–39.
[31] Sorokin LM, Pausch F, Durbeej M, Ekblom P. Differential expression of five laminin alpha (1-5) chains in developing and adult mouse kidney. Dev Dyn 1997;210:446–62.
[32] Sanes JR, Engvall E, Butkowski R, Hunter DD. Molecular heterogeneity of basal laminae: isoforms of laminin and collagen IV at the neuromuscular junction and elsewhere. J Cell Biol 1990;111:1685–99.
[33] Miner JH, Sanes JR. Collagen IV alpha 3, alpha 4, and alpha 5 chains in rodent basal laminae: sequence, distribution, association with laminins, and developmental switches. J Cell Biol 1994;127:879–91.
[34] Pedrosa-Domellof F, Tiger CF, Virtanen I, Thornell LE, Gullberg D. Laminin chains in developing and adult human myotendinous junctions. J Histochem Cytochem 2000;48:201–10.
[35] Noakes PG, Gautam M, Mudd J, Sanes JR, Merlie JP. Aberrant differentiation of neuromuscular junctions in mice lacking s-laminin/laminin beta 2. Nature 1995;374:258–62.
[36] Patton BL, Chiu AY, Sanes JR. Synaptic laminin prevents glial entry into the synaptic cleft. Nature 1998;393:698–701.
[37] Keene DR, Sakai LY, Lunstrum GP, Morris NP, Burgeson RE. Type VII collagen forms an extended network of anchoring fibrils. J Cell Biol 1987;104:611–21.
[38] Rousselle P, Lunstrum GP, Keene DR, Burgeson RE. Kalinin: an epithelium-specific basement membrane adhesion molecule that is a component of anchoring filaments. J Cell Biol 1991;114:567–76.
[39] Rousselle P, Keene DR, Ruggiero F, Champliaud MF, Rest M, Burgeson RE. Laminin 5 binds the NC-1 domain of type VII collagen. J Cell Biol 1997;138:719–28.
[40] Pulkkinen L, Uitto J. Mutation analysis and molecular genetics of epidermolysis bullosa. Matrix Biol 1999;18:29–42.
[41] Champliaud MF, Lunstrum GP, Rousselle P, Nishiyama T, Keene DR, Burgeson RE. Human amnion contains a novel laminin variant, laminin 7, which like laminin 6, covalently associates with laminin 5 to promote stable epithelial-stromal attachment. J Cell Biol 1996;132:1189–98.
[42] Jones JC, Lane K, Hopkinson SB, Lecuona E, Geiger RC, Dean DA, et al. Laminin-6 assembles into multimolecular fibrillar complexes with perlecan and participates in mechanical-signal transduction via a dystroglycan-dependent, integrin-independent mechanism. J Cell Sci 2005;118:2557–66.
[43] Thyboll J, Kortesmaa J, Cao R, Soininen R, Wang L, Iivanainen A, et al. Deletion of the laminin alpha4 chain leads to impaired microvessel maturation. Mol Cell Biol 2002;22:1194–202.
[44] Geberhiwot T, Assefa D, Kortesmaa J, Ingerpuu S, Pedraza C, Wondimu Z, et al. Laminin-8 (alpha4beta1gamma1) is synthesized by lymphoid cells, promotes lymphocyte migration and costimulates T cell proliferation. J Cell Sci 2001;114:423–33.

[45] Niimi T, Kumagai C, Okano M, Kitagawa Y. Differentiation-dependent expression of laminin-8 (alpha 4 beta 1 gamma 1) mRNAs in mouse 3T3-L1 adipocytes. Matrix Biol 1997;16:223–30.

[46] Vaicik MK, Thyboll KJ, Moverare-Skrtic S, Kortesmaa J, Soininen R, Bergstrom G, et al. Laminin alpha4 deficient mice exhibit decreased capacity for adipose tissue expansion and weight gain. PLoS One 2014;9:e109854.

[47] Wallquist W, Plantman S, Thams S, Thyboll J, Kortesmaa J, Lannergren J, et al. Impeded interaction between Schwann cells and axons in the absence of laminin alpha4. J Neurosci 2005;25:3692–700.

[48] Yang D, Bierman J, Tarumi YS, Zhong YP, Rangwala R, Proctor TM, et al. Coordinate control of axon defasciculation and myelination by laminin-2 and -8. J Cell Biol 2005;168:655–66.

[49] Patton BL, Cunningham JM, Thyboll J, Kortesmaa J, Westerblad H, Edstrom L, et al. Properly formed but improperly localized synaptic specializations in the absence of laminin alpha4. Nat Neurosci 2001;4:597–604.

[50] Berfield AK, Hansen KM, Abrass CK. Rat glomerular mesangial cells require laminin-9 to migrate in response to insulin-like growth factor binding protein-5. Am J Physiol Cell Physiol 2006;291:C589–99.

[51] Pouliot N, Saunders NA, Kaur P. Laminin 10/11: an alternative adhesive ligand for epidermal keratinocytes with a functional role in promoting proliferation and migration. Exp Dermatol 2002;11:387–97.

[52] Hayashi K, Mochizuki M, Nomizu M, Uchinuma E, Yamashina S, Kadoya Y. Inhibition of hair follicle growth by a laminin-1 G-domain peptide, RKRLQVQLSIRT, in an organ culture of isolated vibrissa rudiment. J Invest Dermatol 2002;118:712–8.

[53] Nanba D, Hieda Y, Nakanishi Y. Remodeling of desmosomal and hemidesmosomal adhesion systems during early morphogenesis of mouse pelage hair follicles. J Invest Dermatol 2000;114:171–7.

[54] Li J, Tzu J, Chen Y, Zhang YP, Nguyen NT, Gao J, et al. Laminin-10 is crucial for hair morphogenesis. EMBO J 2003;22:2400–10.

[55] DeRouen MC, Zhen H, Tan SH, Williams S, Marinkovich MP, Oro AE. Laminin-511 and integrin beta-1 in hair follicle development and basal cell carcinoma formation. BMC Dev Biol 2010;10:112.

[56] Nguyen NM, Miner JH, Pierce RA, Senior RM. Laminin alpha 5 is required for lobar septation and visceral pleural basement membrane formation in the developing mouse lung. Dev Biol 2002;246:231–44.

[57] Miner JH, Cunningham J, Sanes JR. Roles for laminin in embryogenesis: exencephaly, syndactyly, and placentopathy in mice lacking the laminin alpha5 chain. J Cell Biol 1998;143:1713–23.

[58] Smyth N, Vatansever HS, Murray P, Meyer M, Frie C, Paulsson M, et al. Absence of basement membranes after targeting the LAMC1 gene results in embryonic lethality due to failure of endoderm differentiation. J Cell Biol 1999;144:151–60.

[59] St John PL, Abrahamson DR. Glomerular endothelial cells and podocytes jointly synthesize laminin-1 and -11 chains. Kidney Int 2001;60:1037–46.

[60] Koch M, Olson PF, Albus A, Jin W, Hunter DD, Brunken WJ, et al. Characterization and expression of the laminin gamma3 chain: a novel, non-basement membrane-associated, laminin chain. J Cell Biol 1999;145:605–18.

[61] Egles C, Claudepierre T, Manglapus MK, Champliaud MF, Brunken WJ, Hunter DD. Laminins containing the beta2 chain modulate the precise organization of CNS synapses. Mol Cell Neurosci 2007;34:288–98.

[62] Cohen J, Burne JF, McKinlay C, Winter J. The role of laminin and the laminin/fibronectin receptor complex in the outgrowth of retinal ganglion cell axons. Dev Biol 1987;122:407–18.

[63] Sarthy PV, Fu M. Localization of laminin B1 mRNA in retinal ganglion cells by in situ hybridization. J Cell Biol 1990;110:2099–108.

[64] Dong LJ, Chung AE. The expression of the genes for entactin, laminin A, laminin B1 and laminin B2 in murine lens morphogenesis and eye development. Differentiation 1991;48:157–72.

[65] Morissette N, Carbonetto S. Laminin alpha 2 chain (M chain) is found within the pathway of avian and murine retinal projections. J Neurosci 1995;15:8067–82.

[66] Hunter DD, Murphy MD, Olsson CV, Brunken WJ. S-laminin expression in adult and developing retinae: a potential cue for photoreceptor morphogenesis. Neuron 1992;8:399–413.

[67] Libby RT, Hunter DD, Brunken WJ. Developmental expression of laminin beta 2 in rat retina. Further support for a role in rod morphogenesis. Invest Ophthalmol Vis Sci 1996;37:1651–61.

[68] Hunter DD, Brunken WJ. Beta 2 laminins modulate neuronal phenotype in the rat retina. Mol Cell Neurosci 1997;10:7–15.

[69] Libby RT, Lavallee CR, Balkema GW, Brunken WJ, Hunter DD. Disruption of laminin beta2 chain production causes alterations in morphology and function in the CNS. J Neurosci 1999;19:9399–411.

[70] Libby RT, Champliaud MF, Claudepierre T, Xu Y, Gibbons EP, Koch M, et al. Laminin expression in adult and developing retinae: evidence of two novel CNS laminins. J Neurosci 2000;20:6517–28.

[71] Hunter DD, Shah V, Merlie JP, Sanes JR. A laminin-like adhesive protein concentrated in the synaptic cleft of the neuromuscular junction. Nature 1989;338:229–34.

[72] Aldinger KA, Mosca SJ, Tetreault M, Dempsey JC, Ishak GE, Hartley T, et al. Mutations in LAMA1 cause cerebellar dysplasia and cysts with and without retinal dystrophy. Am J Hum Genet 2014;95:227–34.

[73] Ichikawa-Tomikawa N, Ogawa J, Douet V, Xu Z, Kamikubo Y, Sakurai T, et al. Laminin alpha1 is essential for mouse cerebellar development. Matrix Biol 2012;31:17–28.

[74] Ning L, Kurihara H, de Vega S, Ichikawa-Tomikawa N, Xu Z, Nonaka R, et al. Laminin alpha1 regulates age-related mesangial cell proliferation and mesangial matrix accumulation through the TGF-beta pathway. Am J Pathol 2014;184:1683–94.

[75] Edwards MM, Mammadova-Bach E, Alpy F, Klein A, Hicks WL, Roux M, et al. Mutations in Lama1 disrupt retinal vascular development and inner limiting membrane formation. J Biol Chem 2010;285:7697–711.

[76] Palu E, Liesi P. Differential distribution of laminins in Alzheimer disease and normal human brain tissue. J Neurosci Res 2002;69:243–56.

[77] Gee SH, Blacher RW, Douville PJ, Provost PR, Yurchenco PD, Carbonetto S. Laminin-binding protein 120 from brain is closely related to the dystrophin-associated glycoprotein, dystroglycan, and binds with high affinity to the major heparin binding domain of laminin. J Biol Chem 1993;268:14972–80.

[78] Taraboletti G, Rao CN, Krutzsch HC, Liotta LA, Roberts DD. Sulfatide-binding domain of the laminin A chain. J Biol Chem 1990;265:12253–8.

[79] Miner JH. Developmental biology of glomerular basement membrane components. Curr Opin Nephrol Hypertens 1998;7:13–9.

[80] Scheele S, Falk M, Franzen A, Ellin F, Ferletta M, Lonai P, et al. Laminin alpha1 globular domains 4-5 induce fetal development but are not vital for embryonic basement membrane assembly. Proc Natl Acad Sci USA 2005;102:1502–6.

[81] Abrahamson DR, Isom K, Roach E, Stroganova L, Zelenchuk A, Miner JH, et al. Laminin compensation in collagen alpha3(IV) knockout (Alport) glomeruli contributes to permeability defects. J Am Soc Nephrol 2007;18:2465–72.
[82] Gavassini BF, Carboni N, Nielsen JE, Danielsen ER, Thomsen C, Svenstrup K, et al. Clinical and molecular characterization of limb-girdle muscular dystrophy due to LAMA2 mutations. Muscle Nerve 2011;44:703–9.
[83] Carboni N, Marrosu G, Porcu M, Mateddu A, Solla E, Cocco E, et al. Dilated cardiomyopathy with conduction defects in a patient with partial merosin deficiency due to mutations in the laminin-alpha2-chain gene: a chance association or a novel phenotype? Muscle Nerve 2011;44:826–8.
[84] Carboni N, Sardu C, Cocco E, Marrosu G, Manzi RC, Nissardi V, et al. Cardiac involvement in patients with lamin A/C gene mutations: a cohort observation. Muscle Nerve 2012;46:187–92.
[85] Miyagoe-Suzuki Y, Nakagawa M, Takeda S. Merosin and congenital muscular dystrophy. Microsc Res Tech 2000;48:181–91.
[86] Miyagoe Y, Hanaoka K, Nonaka I, Hayasaka M, Nabeshima Y, Arahata K, et al. Laminin alpha2 chain-null mutant mice by targeted disruption of the Lama2 gene: a new model of merosin (laminin 2)-deficient congenital muscular dystrophy. FEBS Lett 1997;415:33–9.
[87] Kuang W, Xu H, Vachon PH, Liu L, Loechel F, Wewer UM, et al. Merosin-deficient congenital muscular dystrophy. Partial genetic correction in two mouse models. J Clin Invest 1998;102:844–52.
[88] Hager M, Gawlik K, Nystrom A, Sasaki T, Durbeej M. Laminin {alpha}1 chain corrects male infertility caused by absence of laminin {alpha}2 chain. Am J Pathol 2005;167:823–33.
[89] Yuasa K, Fukumoto S, Kamasaki Y, Yamada A, Fukumoto E, Kanaoka K, et al. Laminin alpha2 is essential for odontoblast differentiation regulating dentin sialoprotein expression. J Biol Chem 2004;279:10286–92.
[90] Oliviero P, Chassagne C, Salichon N, Corbier A, Hamon G, Marotte F, et al. Expression of laminin alpha2 chain during normal and pathological growth of myocardium in rat and human. Cardiovasc Res 2000;46:346–55.
[91] Xu H, Wu XR, Wewer UM, Engvall E. Murine muscular dystrophy caused by a mutation in the laminin alpha 2 (Lama2) gene. Nat Genet 1994;8:297–302.
[92] Vidal F, Baudoin C, Miquel C, Galliano MF, Christiano AM, Uitto J, et al. Cloning of the laminin alpha 3 chain gene (LAMA3) and identification of a homozygous deletion in a patient with Herlitz junctional epidermolysis bullosa. Genomics 1995;30:273–80.
[93] Kivirikko S, McGrath JA, Baudoin C, Aberdam D, Ciatti S, Dunnill MG, et al. A homozygous nonsense mutation in the alpha 3 chain gene of laminin 5 (LAMA3) in lethal (Herlitz) junctional epidermolysis bullosa. Hum Mol Genet 1995;4:959–62.
[94] Ryan MC, Lee K, Miyashita Y, Carter WG. Targeted disruption of the LAMA3 gene in mice reveals abnormalities in survival and late stage differentiation of epithelial cells. J Cell Biol 1999;145:1309–23.
[95] Abrass CK, Berfield AK, Ryan MC, Carter WG, Hansen KM. Abnormal development of glomerular endothelial and mesangial cells in mice with targeted disruption of the lama3 gene. Kidney Int 2006;70:1062–71.
[96] Pierce RA, Griffin GL, Mudd MS, Moxley MA, Longmore WJ, Sanes JR, et al. Expression of laminin alpha3, alpha4, and alpha5 chains by alveolar epithelial cells and fibroblasts. Am J Respir Cell Mol Biol 1998;19:237–44.
[97] Amin K, Janson C, Seveus L, Miyazaki K, Virtanen I, Venge P. Uncoordinated production of Laminin-5 chains in airways epithelium of allergic asthmatics. Respir Res 2005;6:110.

[98] Kunneken K, Pohlentz G, Schmidt-Hederich A, Odenthal U, Smyth N, Peter-Katalinic J, et al. Recombinant human laminin-5 domains. Effects of heterotrimerization, proteolytic processing, and N-glycosylation on alpha3beta1 integrin binding. J Biol Chem 2004;279:5184–93.
[99] Goldfinger LE, Jiang L, Hopkinson SB, Stack MS, Jones JC. Spatial regulation and activity modulation of plasmin by high affinity binding to the G domain of the alpha 3 subunit of laminin-5. J Biol Chem 2000;275:34887–93.
[100] Utani A, Nomizu M, Matsuura H, Kato K, Kobayashi T, Takeda U, et al. A unique sequence of the laminin alpha 3 G domain binds to heparin and promotes cell adhesion through syndecan-2 and -4. J Biol Chem 2001;276:28779–88.
[101] Bizama C, Benavente F, Salvatierra E, Gutierrez-Moraga A, Espinoza JA, Fernandez EA, et al. The low-abundance transcriptome reveals novel biomarkers, specific intracellular pathways and targetable genes associated with advanced gastric cancer. Int J Cancer 2014;134:755–64.
[102] Knoll R, Postel R, Wang J, Kratzner R, Hennecke G, Vacaru AM, et al. Laminin-alpha4 and integrin-linked kinase mutations cause human cardiomyopathy via simultaneous defects in cardiomyocytes and endothelial cells. Circulation 2007;116:515–25.
[103] Takkunen M, Ainola M, Vainionpaa N, Grenman R, Patarroyo M, Garcia de HA, et al. Epithelial-mesenchymal transition downregulates laminin alpha5 chain and upregulates laminin alpha4 chain in oral squamous carcinoma cells. Histochem Cell Biol 2008;130:509–25.
[104] Vainionpaa N, Lehto VP, Tryggvason K, Virtanen I. Alpha4 chain laminins are widely expressed in renal cell carcinomas and have a de-adhesive function. Lab Invest 2007;87:780–91.
[105] Lian J, Dai X, Li X, He F. Identification of an active site on the laminin alpha4 chain globular domain that binds to alphavbeta3 integrin and promotes angiogenesis. Biochem Biophys Res Commun 2006;347:248–53.
[106] Matsuura H, Momota Y, Murata K, Matsushima H, Suzuki N, Nomizu M, et al. Localization of the laminin alpha4 chain in the skin and identification of a heparin-dependent cell adhesion site within the laminin alpha4 chain C-terminal LG4 module. J Invest Dermatol 2004;122:614–20.
[107] Nagato S, Nakagawa K, Harada H, Kohno S, Fujiwara H, Sekiguchi K, et al. Downregulation of laminin alpha4 chain expression inhibits glioma invasion in vitro and in vivo. Int J Cancer 2005;117:41–50.
[108] Huang X, Ji G, Wu Y, Wan B, Yu L. LAMA4, highly expressed in human hepatocellular carcinoma from Chinese patients, is a novel marker of tumor invasion and metastasis. J Cancer Res Clin Oncol 2008;134:705–14.
[109] Miner JH, Li C. Defective glomerulogenesis in the absence of laminin alpha5 demonstrates a developmental role for the kidney glomerular basement membrane. Dev Biol 2000;217:278–89.
[110] Fukumoto S, Miner JH, Ida H, Fukumoto E, Yuasa K, Miyazaki H, et al. Laminin alpha5 is required for dental epithelium growth and polarity and the development of tooth bud and shape. J Biol Chem 2006;281:5008–16.
[111] Rebustini IT, Patel VN, Stewart JS, Layvey A, Georges-Labouesse E, Miner JH, et al. Laminin alpha5 is necessary for submandibular gland epithelial morphogenesis and influences FGFR expression through beta1 integrin signaling. Dev Biol 2007;308:15–29.
[112] Kikkawa Y, Miner JH. Molecular dissection of laminin alpha 5 in vivo reveals separable domain-specific roles in embryonic development and kidney function. Dev Biol 2006;296:265–77.
[113] Bair EL, Chen ML, McDaniel K, Sekiguchi K, Cress AE, Nagle RB, et al. Membrane type 1 matrix metalloprotease cleaves laminin-10 and promotes prostate cancer cell migration. Neoplasia 2005;7:380–9.

[114] Kikkawa Y, Virtanen I, Miner JH. Mesangial cells organize the glomerular capillaries by adhering to the G domain of laminin alpha5 in the glomerular basement membrane. J Cell Biol 2003;161:187–96.
[115] Miner JH. Laminins and their roles in mammals. Microsc Res Tech 2008;71:349–56.
[116] Goldberg S, Adair-Kirk TL, Senior RM, Miner JH. Maintenance of glomerular filtration barrier integrity requires laminin alpha5. J Am Soc Nephrol 2010;21:579–86.
[117] Shannon MB, Patton BL, Harvey SJ, Miner JH. A hypomorphic mutation in the mouse laminin alpha5 gene causes polycystic kidney disease. J Am Soc Nephrol 2006;17:1913–22.
[118] Radmanesh F, Caglayan AO, Silhavy JL, Yilmaz C, Cantagrel V, Omar T, et al. Mutations in LAMB1 cause cobblestone brain malformation without muscular or ocular abnormalities. Am J Hum Genet 2013;92:468–74.
[119] Wu Y, Wang XF, Mo XA, Li JM, Yuan J, Zheng JO, et al. Expression of laminin beta1 and integrin alpha2 in the anterior temporal neocortex tissue of patients with intractable epilepsy. Int J Neurosci 2011;121:323–8.
[120] Zenker M, Aigner T, Wendler O, Tralau T, Muntefering H, Fenski R, et al. Human laminin beta2 deficiency causes congenital nephrosis with mesangial sclerosis and distinct eye abnormalities. Hum Mol Genet 2004;13:2625–32.
[121] Hasselbacher K, Wiggins RC, Matejas V, Hinkes BG, Mucha B, Hoskins BE, et al. Recessive missense mutations in LAMB2 expand the clinical spectrum of LAMB2-associated disorders. Kidney Int 2006;70:1008–12.
[122] Noakes PG, Miner JH, Gautam M, Cunningham JM, Sanes JR, Merlie JP. The renal glomerulus of mice lacking s-laminin/laminin beta 2: nephrosis despite molecular compensation by laminin beta 1. Nat Genet 1995;10:400–6.
[123] Ooba T, Ishikawa T, Yamaguchi K, Kondo Y, Sakamoto Y, Fujisawa M. Expression and distribution of laminin chains in the testis for patients with azoospermia. J Androl 2008;29:147–52.
[124] Taniguchi Y, Ido H, Sanzen N, Hayashi M, Sato-Nishiuchi R, Futaki S, et al. The C-terminal region of laminin beta chains modulates the integrin binding affinities of laminins. J Biol Chem 2009;284:7820–31.
[125] Mayer U, Nischt R, Poschl E, Mann K, Fukuda K, Gerl M, et al. A single EGF-like motif of laminin is responsible for high affinity nidogen binding. EMBO J 1993;12:1879–85.
[126] Pulkkinen L, Meneguzzi G, McGrath JA, Xu Y, Blanchet-Bardon C, Ortonne JP, et al. Predominance of the recurrent mutation R635X in the LAMB3 gene in European patients with Herlitz junctional epidermolysis bullosa has implications for mutation detection strategy. J Invest Dermatol 1997;109:232–7.
[127] Mellerio JE, Eady RA, Atherton DJ, Lake BD, McGrath JA. E210K mutation in the gene encoding the beta3 chain of laminin-5 (LAMB3) is predictive of a phenotype of generalized atrophic benign epidermolysis bullosa. Br J Dermatol 1998;139:325–31.
[128] Pasmooij AM, Pas HH, Bolling MC, Jonkman MF. Revertant mosaicism in junctional epidermolysis bullosa due to multiple correcting second-site mutations in LAMB3. J Clin Invest 2007;117:1240–8.
[129] Floeth M, Bruckner-Tuderman L. Digenic junctional epidermolysis bullosa: mutations in COL17A1 and LAMB3 genes. Am J Hum Genet 1999;65:1530–7.
[130] Kuster JE, Guarnieri MH, Ault JG, Flaherty L, Swiatek PJ. IAP insertion in the murine LamB3 gene results in junctional epidermolysis bullosa. Mamm Genome 1997;8:673–81.
[131] Muhle C, Neuner A, Park J, Pacho F, Jiang Q, Waddington SN, et al. Evaluation of prenatal intra-amniotic LAMB3 gene delivery in a mouse model of Herlitz disease. Gene Ther 2006;13:1665–76.

[132] Sonnenberg A, de Melker AA, Martinez de Velasco AM, Janssen H, Calafat J, Niessen CM. Formation of hemidesmosomes in cells of a transformed murine mammary tumor cell line and mechanisms involved in adherence of these cells to laminin and kalinin. J Cell Sci 1993;106(Pt 4):1083–102.
[133] Oka T, Yamamoto H, Sasaki S, Ii M, Hizaki K, Taniguchi H, et al. Overexpression of beta3/gamma2 chains of laminin-5 and MMP7 in biliary cancer. World J Gastroenterol 2009;15:3865–73.
[134] Wang XM, Li J, Yan MX, Liu L, Jia DS, Geng Q, et al. Integrative analyses identify osteopontin, LAMB3 and ITGB1 as critical pro-metastatic genes for lung cancer. PLoS One 2013;8:e55714.
[135] Udayakumar TS, Chen ML, Bair EL, Von B, Cress AE, Nagle RB, et al. Membrane type-1-matrix metalloproteinase expressed by prostate carcinoma cells cleaves human laminin-5 beta3 chain and induces cell migration. Cancer Res 2003;63:2292–9.
[136] Sercu S, Lambeir AM, Steenackers E, El GA, Geentjens K, Sasaki T, et al. ECM1 interacts with fibulin-3 and the beta 3 chain of laminin 332 through its serum albumin subdomain-like 2 domain. Matrix Biol 2009;28:160–9.
[137] Hammersen J, Hou J, Wunsche S, Brenner S, Winkler T, Schneider H. A new mouse model of junctional epidermolysis bullosa: the LAMB3 628G>A knockin mouse. J Invest Dermatol 2015;135:921–4.
[138] Imura J, Uchida Y, Nomoto K, Ichikawa K, Tomita S, Iijima T, et al. Laminin-5 is a biomarker of invasiveness in cervical adenocarcinoma. Diagn Pathol 2012;7:105.
[139] Nikolova G, Lee H, Berkovitz S, Nelson S, Sinsheimer J, Vilain E, et al. Sequence variant in the laminin gamma1 (LAMC1) gene associated with familial pelvic organ prolapse. Hum Genet 2007;120:847–56.
[140] Ke HL, Ke RH, Li B, Wang XH, Wang YN, Wang XQ. Association between laminin gamma1 expression and meningioma grade, recurrence, and progression-free survival. Acta Neurochir (Wien) 2013;155:165–71.
[141] Kuhn E, Kurman RJ, Soslow RA, Han G, Sehdev AS, Morin PJ, et al. The diagnostic and biological implications of laminin expression in serous tubal intraepithelial carcinoma. Am J Surg Pathol 2012;36:1826–34.
[142] Palu E, Liesi P. Differential distribution of laminins in Alzheimer disease and normal human brain tissue. J Neurosci Res 2002;69:243–56.
[143] Wiksten M, Liebkind R, Laatikainen T, Liesi P. Gamma 1 laminin and its biologically active KDI-domain may guide axons in the floor plate of human embryonic spinal cord. J Neurosci Res 2003;71:338–52.
[144] Patel TR, Morris GA, Zwolanek D, Keene DR, Li J, Harding SE, et al. Nano-structure of the laminin gamma-1 short arm reveals an extended and curved multidomain assembly. Matrix Biol 2010;29:565–72.
[145] Willem M, Miosge N, Halfter W, Smyth N, Jannetti I, Burghart E, et al. Specific ablation of the nidogen-binding site in the laminin gamma1 chain interferes with kidney and lung development. Development 2002;129:2711–22.
[146] Kohfeldt E, Sasaki T, Gohring W, Timpl R. Nidogen-2: a new basement membrane protein with diverse binding properties. J Mol Biol 1998;282:99–109.
[147] Chen ZL, Strickland S. Laminin gamma1 is critical for Schwann cell differentiation, axon myelination, and regeneration in the peripheral nerve. J Cell Biol 2003;163:889–99.
[148] Meng X, Klement JF, Leperi DA, Birk DE, Sasaki T, Timpl R, et al. Targeted inactivation of murine laminin gamma2-chain gene recapitulates human junctional epidermolysis bullosa. J Invest Dermatol 2003;121:720–31.

[149] Nguyen NM, Pulkkinen L, Schlueter JA, Meneguzzi G, Uitto J, Senior RM. Lung development in laminin gamma2 deficiency: abnormal tracheal hemidesmosomes with normal branching morphogenesis and epithelial differentiation. Respir Res 2006;7:28.
[150] Xue LY, Zou SM, Zheng S, Liu XY, Wen P, Yuan YL, et al. Expressions of the gamma2 chain of laminin-5 and secreted protein acidic and rich in cysteine in esophageal squamous cell carcinoma and their relation to prognosis. Chin J Cancer 2011;30:69–78.
[151] Katayama M, Ishizaka A, Sakamoto M, Fujishima S, Sekiguchi K, Asano K, et al. Laminin gamma2 fragments are increased in the circulation of patients with early phase acute lung injury. Intensive Care Med 2010;36:479–86.
[152] Giannelli G, Falk-Marzillier J, Schiraldi O, Stetler-Stevenson WG, Quaranta V. Induction of cell migration by matrix metalloprotease-2 cleavage of laminin-5. Science 1997;277:225–8.
[153] Koshikawa N, Giannelli G, Cirulli V, Miyazaki K, Quaranta V. Role of cell surface metalloprotease MT1-MMP in epithelial cell migration over laminin-5. J Cell Biol 2000;148:615–24.
[154] Amano S, Scott IC, Takahara K, Koch M, Champliaud MF, Gerecke DR, et al. Bone morphogenetic protein 1 is an extracellular processing enzyme of the laminin 5 gamma 2 chain. J Biol Chem 2000;275:22728–35.
[155] Ogawa T, Tsubota Y, Hashimoto J, Kariya Y, Miyazaki K. The short arm of laminin gamma2 chain of laminin-5 (laminin-332) binds syndecan-1 and regulates cellular adhesion and migration by suppressing phosphorylation of integrin beta4 chain. Mol Biol Cell 2007;18:1621–33.
[156] Bubier JA, Sproule TJ, Alley LM, Webb CM, Fine JD, Roopenian DC, et al. A mouse model of generalized non-Herlitz junctional epidermolysis bullosa. J Invest Dermatol 2010;130:1819–28.
[157] Kosanam H, Prassas I, Chrystoja CC, Soleas I, Chan A, Dimitromanolakis A, et al. Laminin, gamma 2 (LAMC2): a promising new putative pancreatic cancer biomarker identified by proteomic analysis of pancreatic adenocarcinoma tissues. Mol Cell Proteomics 2013;12:2820–32.
[158] Smith SC, Nicholson B, Nitz M, Frierson Jr HF, Smolkin M, Hampton G, et al. Profiling bladder cancer organ site-specific metastasis identifies LAMC2 as a novel biomarker of hematogenous dissemination. Am J Pathol 2009;174:371–9.
[159] Barak T, Kwan KY, Louvi A, Demirbilek V, Saygi S, Tuysuz B, et al. Recessive LAMC3 mutations cause malformations of occipital cortical development. Nat Genet 2011;43:590–4.
[160] Denes V, Witkovsky P, Koch M, Hunter DD, Pinzon-Duarte G, Brunken WJ. Laminin deficits induce alterations in the development of dopaminergic neurons in the mouse retina. Vis Neurosci 2007;24:549–62.
[161] Ido H, Ito S, Taniguchi Y, Hayashi M, Sato-Nishiuchi R, Sanzen N, et al. Laminin isoforms containing the gamma3 chain are unable to bind to integrins due to the absence of the glutamic acid residue conserved in the C-terminal regions of the gamma1 and gamma2 chains. J Biol Chem 2008;283:28149–57.
[162] Gersdorff N, Kohfeldt E, Sasaki T, Timpl R, Miosge N. Laminin gamma3 chain binds to nidogen and is located in murine basement membranes. J Biol Chem 2005;280:22146–53.
[163] Radner S, Banos C, Bachay G, Li YN, Hunter DD, Brunken WJ, et al. beta2 and gamma3 laminins are critical cortical basement membrane components: ablation of Lamb2 and Lamc3 genes disrupts cortical lamination and produces dysplasia. Dev Neurobiol 2013;73:209–29.

Chapter 30

Elastin

J.H. Kristensen[1,2], M.A. Karsdal[1]
[1]Nordic Bioscience, Herlev, Denmark; [2]The Technical University of Denmark, Kongens Lyngby, Denmark

Chapter Outline

SUMMARY
Elastin is a key extracellular matrix (ECM) protein that provides resilience and elasticity to tissues and organs. Elastin is roughly 1000 times more flexible than collagens; thus, the main function of elastin is the elasticity of tissues. It is the dominant protein in extensible tissues and is primarily present in the lungs, aorta, and skin. Mutations in the elastin gene may lead to diseases such as Williams–Beuren syndrome, cutis laxa, and supravalvular aortic stenosis (SVAS). The precursor of elastin is tropoelastin. Tropoelastin is derived from fibroblasts, smooth muscle cells, chondrocytes, or endothelial cells before it is processed to elastin by cleavage of its signal peptide. Elastin monomers are crosslinked during the formation of desmosine molecules. Several important and well-described biomarkers of elastin degradation are available.

Elastin is a key ECM protein that provides resilience and elasticity to tissues and organs. It is one of the dominant proteins in extensible tissues and is primarily present in the lungs, aorta, and skin [1]. Elastin constitutes the majority of elastic fibers. During fiber formation, it binds to other structural ECM molecules such as fibulins, and fibrillins with an ε-(γ-glutamyl) lysine bond [2–4]. The binding is facilitated by transglutaminase (Tgase).

The precursor of elastin is tropoelastin. Tropoelastin is derived from fibroblasts, smooth muscle cells, chondrocytes, or endothelial cells before it is processed to elastin by cleavage of its signal peptide [5]. Elastin monomers can be crosslinked during the formation of desmosine molecules that are formed by the crosslinking of three allysine and one lysine residue. The process is facilitated by lysyl oxidase that oxidizes elastin lysine residues into allysine [6]. Elastin contains several hydrophobic domains that are dominated by the residues valine, glycine, and proline (UniProt: P15502 [7]). In addition to the hydrophobic domains elastin has several special domains such as multiple hydroxyproline and allysine sites, a polyalanine domain (residues 97–102), and alanine-rich domains (residues 236–742) as well as the rarely expressed exon 26A [7–10]. Its crosslinks render elastin remarkably hydrophobic and stable, with a long half-life of up to 40 years [11].

The roles of elastin in the development of the lungs in particular have been thoroughly investigated (Table 30.1). Studies in mice have shown that elastin null mutations result in fewer, larger air sacs in the lungs and severe branching defects [12]. Elastin-deficient mice are also more susceptible to smoking-induced emphysema than mice with normal elastin levels [13]. Elastin is also responsible for the elasticity of the conduit arteries [14]. As the body ages, the elastin contents of the arteries declines, thereby increasing arterial stiffness and thus the pulse pressure [14]. This has also been demonstrated in animal models with induced hypertension where the elastin-to-collagen ratio decreases, which results in stiffer arteries [15,16]. For comparison, it is estimated that collagens are more than 1000 times stiffer than elastin [17]. Last, mutations in the elastin gene may lead to diseases such as Williams–Beuren syndrome, cutis laxa, and SVAS [18–20].

Elastin synthesis peaks during the embryonic phase and early childhood [21]. The lower elastin turnover occurring during adulthood enables the detection and quantification of abnormal elastin degradation or distribution [22,23]. Elevated levels of elastin fragments have been found in adult patients diagnosed with chronic obstructive pulmonary disease (COPD), idiopathic pulmonary fibrosis, lung cancer, or cardiovascular diseases [22,24–26]. Furthermore, increased elastin degradation rates, as demonstrated by the presence of increasing matrix metalloproteinase–and human neutrophil elastase–mediated elastin fragmentation, may indicate acute exacerbations in COPD patients [27].

Ordered distribution of elastin depends on controlled elastin crosslinking activity. Circulating elastin fragments containing desmosine can be quantified with a wide range of techniques including enzyme-linked immunosorbent assay, high-performance liquid chromatography–mass spectrometry, and capillary electrophoresis [28]. Levels of these desmosine-containing fragments may be used to monitor fibrosis in interventional studies, although more research is required to confirm the utility of this potential marker. Elastin fragments are not merely a consequence of tissue destruction; they may also act as pathological mediators. In emphysema, elastin fragments stimulate an immune response in lung tissues, leading to the upregulation of cytokines with a further increase in protease expression and subsequently to progressive destruction of the lung ECM [29]. During skin repair, elastin signaling attenuates wound contraction and stimulates protease production, cell migration, and proliferation as well as matrix synthesis [30]. Last, peptides expressed from the rarely transcribed or translated exon 26A are believed to induce cell migration and to have chemotactic properties [10]. Future studies may reveal additional regulatory properties of elastin.

TABLE 30.1 Elastin

Elastin	Description	References
Gene name and number	Gene name ELN, location 7q11.1-q21.1	[31]
Mutations with diseases in humans	Williams–Beuren syndrome, cutis laxa, SVAS	[18–20]
Null mutation in mice	SVAS. Impaired lung development. Mice with elastin deficiencies are more prone to emphysema. Elastin knockout (ELN–/–) mice die shortly after birth.	[12,13,32]
Tissue distribution in healthy states	Primarily in the aorta, skin, and lungs	[1,33]
Tissue distribution in pathologically affected states	Proteolytic degradation and chemical alterations of elastin have been observed in pathologies affecting the lungs and arteries	[16,22,24–27]
Special domains	Multiple hydroxyproline and allysine sites. Polyalanine domain (residues 97–102), alanine-rich domain (residues 236–742). Hydrophobic regions. Exon 26A.	[7–10]
Special neoepitopes	ELM and ELM2 are elastin fragments generated by MMP-9 and MMP-12 cleavage of elastin. ELM7 and EL-NE are elastin fragments generated by MMP-7 and HNE cleavage of elastin, respectively. Upregulation of elastin fragments in the circulation can be coupled with pulmonary or arterial disorders. Peptides derived from exon 26A.	[10,22,25,26]
Protein structure and function	Cleavage of tropoelastin generates elastin monomers. Elastin and tropoelastin consist of hydrophobic and hydrophilic domains. The hydrophobic domains of elastin and crosslinking of elastin monomers render elastic fibers insoluble and stable. Elastin ensures tissue elasticity.	[1,11]
Binding proteins	Binds mainly to other elastin monomers, fibrillins, and fibulins. Binding and crosslinking is facilitated by LOX and Tgase.	[2–4]
Known central function	Structural protein that provides elasticity to tissue. Fragments of elastin may stimulate the immune system.	[12,29]
Animals models with protein affected	Elastin-deficient mice are used to study lung development and susceptibility to emphysema caused by smoking.	[13]
Biomarkers	ELM, ELM2, ELM7, EL-NE, desmosine	[22,23,25,26]

ELN, elastin; *SVAS*, supravalvular aortic stenosis; *MMP*, matrix metalloproteinase; *HNE*, human neutrophil elastase; *LOX*, lysyl oxidase; *Tgase*, transglutaminase.

REFERENCES

[1] Wise SG, Weiss AS. Tropoelastin. Int J Biochem Cell Biol March 2009;41(3):494–7.

[2] Rock MJ, Cain SA, Freeman LJ, Morgan A, Mellody K, Marson A, et al. Molecular basis of elastic fiber formation. Critical interactions and a tropoelastin-fibrillin-1 cross-link. J Biol Chem May 28, 2004;279(22):23748–58.

[3] Chapman SL, Sicot FX, Davis EC, Huang J, Sasaki T, Chu ML, et al. Fibulin-2 and fibulin-5 cooperatively function to form the internal elastic lamina and protect from vascular injury. Arterioscler Thromb Vasc Biol January 2010;30(1):68–74.

[4] Kothapalli CR, Ramamurthi A. Lysyl oxidase enhances elastin synthesis and matrix formation by vascular smooth muscle cells. J Tissue Eng Regen Med December 2009;3(8):655–61.

[5] Starcher BC. Elastin and the lung. Thorax August 1986;41(8):577–85.

[6] Csiszar K. Lysyl oxidases: a novel multifunctional amine oxidase family. Prog Nucleic Acid Res Mol Biol 2001;70:1–32.

[7] Combet C, Blanchet C, Geourjon C, Deleage G. NPS@: network protein sequence analysis. Trends Biochem Sci March 2000;25(3):147–50.

[8] Schmelzer CE, Getie M, Neubert RH. Mass spectrometric characterization of human skin elastin peptides produced by proteolytic digestion with pepsin and thermitase. J Chromatogr A August 12, 2005;1083(1–2):120–6.

[9] Dyksterhuis LB, Weiss AS. Homology models for domains 21-23 of human tropoelastin shed light on lysine crosslinking. Biochem Biophys Res Commun June 11, 2010;396(4):870–3.

[10] Bisaccia F, Castiglione-Morelli MA, Spisani S, Ostuni A, Serafini-Fracassini A, Bavoso A, et al. The amino acid sequence coded by the rarely expressed exon 26A of human elastin contains a stable beta-turn with chemotactic activity for monocytes. Biochemistry August 4, 1998;37(31):11128–35.

[11] Arribas SM, Hinek A, Gonzalez MC. Elastic fibres and vascular structure in hypertension. Pharmacol Ther September 2006;111(3):771–91.

[12] Wendel DP, Taylor DG, Albertine KH, Keating MT, Li DY. Impaired distal airway development in mice lacking elastin. Am J Respir Cell Mol Biol September 2000;23(3):320–6.

[13] Shifren A, Durmowicz AG, Knutsen RH, Hirano E, Mecham RP. Elastin protein levels are a vital modifier affecting normal lung development and susceptibility to emphysema. Am J Physiol Lung Cell Mol Physiol March 2007;292(3):L778–87.

[14] Greenwald SE. Ageing of the conduit arteries. J Pathol January 2007;211(2):157–72.

[15] Wolinsky H. Response of the rat aortic media to hypertension. Morphological and chemical studies. Circ Res April 1970;26(4):507–22.

[16] Wagenseil JE, Mecham RP. Elastin in large artery stiffness and hypertension. J Cardiovasc Transl Res June 2012;5(3):264–73.

[17] Shadwick RE. Mechanical design in arteries. J Exp Biol December 1999;202(Pt 23):3305–13.

[18] Delio M, Pope K, Wang T, Samanich J, Haldeman-Englert CR, Kaplan P, et al. Spectrum of elastin sequence variants and cardiovascular phenotypes in 49 patients with Williams-Beuren syndrome. Am J Med Genet A March 2013;161A(3):527–33.

[19] Callewaert B, Renard M, Hucthagowder V, Albrecht B, Hausser I, Blair E, et al. New insights into the pathogenesis of autosomal-dominant cutis laxa with report of five ELN mutations. Hum Mutat April 2011;32(4):445–55.

[20] Curran ME, Atkinson DL, Ewart AK, Morris CA, Leppert MF, Keating MT. The elastin gene is disrupted by a translocation associated with supravalvular aortic stenosis. Cell April 9, 1993;73(1):159–68.

[21] Swee MH, Parks WC, Pierce RA. Developmental regulation of elastin production. Expression of tropoelastin pre-mRNA persists after down-regulation of steady-state mRNA levels. J Biol Chem June 23, 1995;270(25):14899–906.

[22] Skjot-Arkil H, Clausen RE, Rasmussen LM, Wang W, Wang Y, Zheng Q, et al. Acute myocardial infarction and pulmonary diseases result in two different degradation profiles of elastin as quantified by two novel ELISAs. PLoS One 2013;8(6):e60936.

[23] Lindberg CA, Engstrom G, de Verdier MG, Nihlen U, Anderson M, Forsman-Semb K, et al. Total desmosines in plasma and urine correlate with lung function. Eur Respir J April 2012;39(4):839–45.

[24] Skjot-Arkil H, Clausen RE, Nguyen QH, Wang Y, Zheng Q, Martinez FJ, et al. Measurement of MMP-9 and -12 degraded elastin (ELM) provides unique information on lung tissue degradation. BMC Pulm Med 2012;12(1):34.

[25] Kristensen JH, Karsdal MA, Sand JM, Willumsen N, Diefenbach C, Svensson B, et al. Serological assessment of neutrophil elastase activity on elastin during lung ECM remodeling. BMC Pulm Med 2015;15(1):53.

[26] Kristensen JH, Larsen L, Dasgupta B, Brodmerkel C, Curran M, Karsdal MA, et al. Levels of circulating MMP-7 degraded elastin are elevated in pulmonary disorders. Clin Biochem November 2015;48(16-17):1083–8.

[27] Sand JM, Knox AJ, Lange P, Sun S, Kristensen JH, Leeming DJ, et al. Accelerated extracellular matrix turnover during exacerbations of COPD. Respir Res June 11, 2015;16(1):69.

[28] Luisetti M, Ma S, Iadarola P, Stone PJ, Viglio S, Casado B, et al. Desmosine as a biomarker of elastin degradation in COPD: current status and future directions. Eur Respir J November 2008;32(5):1146–57.

[29] Lee SH, Goswami S, Grudo A, Song LZ, Bandi V, Goodnight-White S, et al. Antielastin autoimmunity in tobacco smoking-induced emphysema. Nat Med May 2007;13(5):567–9.

[30] Almine JF, Wise SG, Weiss AS. Elastin signaling in wound repair. Birth Defects Res C Embryo Today September 2012;96(3):248–57.

[31] Fazio MJ, Mattei MG, Passage E, Chu ML, Black D, Solomon E, et al. Human elastin gene: new evidence for localization to the long arm of chromosome 7. Am J Hum Genet April 1991;48(4):696–703.

[32] Li DY, Brooke B, Davis EC, Mecham RP, Sorensen LK, Boak BB, et al. Elastin is an essential determinant of arterial morphogenesis. Nature May 21, 1998;393(6682):276–80.

[33] Chrzanowski P, Keller S, Cerreta J, Mandl I, Turino GM. Elastin content of normal and emphysematous lung parenchyma. Am J Med September 1980;69(3):351–9.

Chapter 31

Structural Biomarkers

A.C. Bay-Jensen, J.M.B. Sand, F. Genovese, A.S. Siebuhr, M.J. Nielsen, D.J. Leeming, T. Manon-Jensen, M.A. Karsdal
Nordic Bioscience, Herlev, Denmark

Chapter Outline

SUMMARY

A central event in tissue remodeling is expression and activation of extracellular matrix (ECM)–degrading proteinases and collagenases, resulting in the release of small protein fragments called neoepitopes into the circulation. These neoepitopes can and are being targeted as biochemical markers of "end products of tissue destruction."

Neoepitopes can include different posttranslational modifications (PTMs) such as citrullination, nitrosylation, glycosylation, and isomerization. Each modification is a result from specific and local physiological processes. Identification of PTMs on neoepitopes may deliver unique disease-specific biomarkers and have been identified in musculoskeletal, cardiovascular, fibrotic, and neurodegenerative diseases. The neoepitope biomarkers relying on analytes that are modified by multiple PTMs may become optimal tools that meet the burden of disease, investigatory, prognostic, efficacy of intervention and diagnosis (BIPED) biochemical marker "usefulness" criteria.

In this chapter, the use of structural biomarkers is discussed, with emphasis on various PTMs in different disease indications. The use of the BIPED criteria is likewise discussed, as well as the importance of understanding the effect of the matrix in which the biomarker is measured, such as serum, urine, and other body fluids, on interpretation of results.

ECM remodeling, in which old or damaged proteins are degraded and replaced by new, intact proteins, is a key process in tissue homeostasis. Specific proteolytic activities are a prerequisite for a range of cellular functions and interactions within the ECM during this remodeling. These specific activities are balanced to promote tissue turnover and renewal. In disease states, this repair–response balance is disrupted, leading to excessive tissue turnover in which the ECM proteins are replaced by new and potentially different ECM proteins. An imbalance in ECM remodeling results in the release of small ECM protein fragments into the systemic circulation. If detected, these fragments may be used as molecular biochemical markers of various pathologies.

Several biochemical markers (biomarkers) measuring ECM remodeling have become available in recent decades, and they may provide tools for investigation of the ECM imbalance observed in different diseases as well as providing in vitro diagnostic tools for the assessment of treatment efficacy and prognosis of patients diagnosed with chronic diseases such as osteoporosis (OP) [1,2], rheumatoid arthritis (RA), chronic obstructive pulmonary disease, and cancers [3]. Many of the biomarkers have been categorized according to the BIPED classification developed by the National Institutes of Health–industry partnership funded by the Osteoarthritis Biochemical Markers Network [4–7]. The Food and Drug Administration critical path initiative launched in 2004 further emphasized the need for translational science and the use of biochemical markers during drug discovery and development [3].

Below, we describe the biomarkers that are available to measure ECM turnover.

THE CONVERGENCE OF DIFFERENT PATHWAYS LEADS TO ALTERED TURNOVER OF THE ECM

A common denominator of many chronic and inflammatory diseases is elevated levels of proinflammatory cytokines, including, but not limited to, the interleukin (IL)-1, IL-6, IL-17, and IL-18 and tumor necrosis factor (TNF)-α [8]. The combined proinflammatory burden is associated with substantial cell surface receptor activation in different tissues, resulting in substantial alterations in cellular signaling mechanisms involving, in particular, nonreceptor tyrosine kinases such as Syk, Jak, Src, and PI3K [9]. This leads to increased tissue turnover with an imbalance between tissue formation and degradation, resulting in massive connective tissue destruction and failure [10]. Such antiinflammatory treatments have focused on reducing or ablating the pathological activity in tissues, either by directly targeting the pathological cytokines (as is the case with antibodies against cytokines or their receptors), or by reducing the activity of their downstream signaling mediators (so-called kinase inhibitors). The kinase signaling pathways are prime targets for pharmaceutical intervention and have proven successful in treating inflammation [11].

One example of successful targeting of the kinase signaling pathway is in RA, a chronic autoimmune disease characterized by inflammation in multiple synovial joints leading to tissue destruction and joint failure. In principle, the inflammatory burden of disease results in the activation of transmembrane cytokine receptors leading to the activation of intracellular signaling pathways including JNK, ITAM, PLC, NFAT, AKT, STAT, ERK, SYK, LYN, FYN, and JAK [11–13]. Activation of the signaling pathways ultimately leads to the expression of powerful proteolytic and collagenolytic enzymes, in particular the matrix metalloproteinases (MMPs) and aggrecanases [14,15] (Fig. 31.1). The proteolytic enzymes are secreted partly by cells of the affected tissues and partly by elevated levels of invading inflammatory cells such as macrophages, T cells, and neutrophils. This increased proteolytic activity destroys the ECM, resulting in the release of ECM fragments that are cleared from the joint and released into the circulation as biomarkers [16–18]. Furthermore, the inflammatory burden activates the fibroblasts to become myofibroblasts that express novel ECM proteins. In this process, proteins fragments are also released into the circulation from different types of collagen, many of which contain propeptides that are posttranslationally cleaved as part of tissue incorporation [19]. In RA, MMP-mediated degradation of ECM collagen proteins has resulted in the

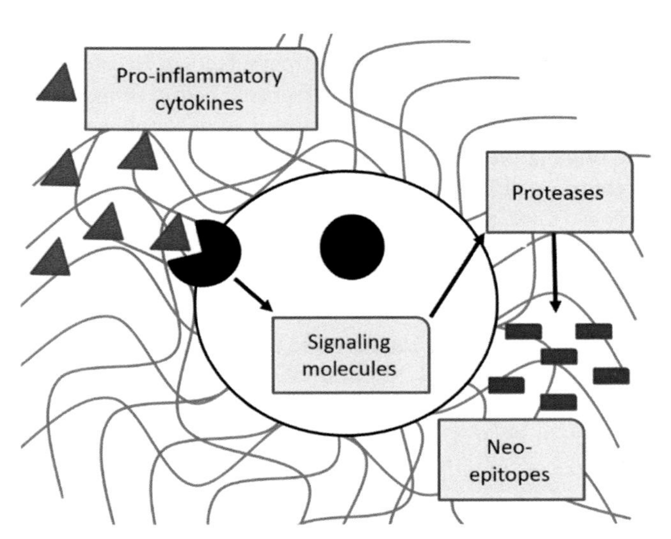

FIGURE 31.1 Neoepitopes of the extracellular matrix (ECM) are the end product of tissue destruction. Damage to the tissue increases the level of proinflammatory cytokines that initiate a signaling cascade leading to an upregulation of proteases such as matrix metalloproteinases that degrade the ECM, resulting in the release of neoepitopes that can be used as biomarkers.

development of specific biomarkers: C1M and ICTP; C2M, C2C, C-terminal telopeptide of type II collagen (CTX-II), and C3M, measuring degradation of types I, II, and III collagen, respectively [20–23]. Other noteworthy biomarkers of the joint are: CTX-I, used to measure bone resorption mediated by cathepsin K activity [24]; ARGS or AGNx1 and FFGV, measuring aggrecan degraded by aggrecanases or MMPs [25,26]; and PIINP, Pro-C2, type IIA procollagen amino-terminal propeptide (PIIANP), formation biomarkers of type II collagen degradation [27–31]. As a tissue-specific example, type II collagen is almost exclusively expressed in cartilage. C2M, CTX-II, and C2C are consequently markers of cartilage degradation; hence, they are used as a tool to track changes in cartilage turnover in rheumatological conditions [3].

In RA and other chronic inflammatory diseases a range of cytokines may activate different cellular pathways, which end in the production of the "usual suspects" of a battery of proteases inducing tissue turnover. Although the cytokine signals may be many, and the intracellular signaling pathways diverse, the end products that are specific for given tissues—the small protein fragments that are released from the tissue due to excessive protease activity—may be considered the result of the convergence of pathways.

DIFFERENT CLASSES OF ECM BIOMARKERS

ECM biomarkers can be divided into three classes (Fig. 31.2). The first class (A) is often based on total protein measurement, and different prerequisites are needed for each assessment (Fig. 31.2A). The second class (B) assesses the presence of a single specific ECM protein fragment or PTMs of an ECM protein. The third class (C) of biomarkers enables measurement of multiple PTMs (eg, those arising from fragmentation together with oxidation; see description below). Because we now know that collagens and other ECM proteins are degraded before being released into the circulation, the assay developed to measure the proteins should be designed so that a most of the original protein components can be detected. In an enzyme-linked immunosorbent assay (ELISA), this could be achieved by using two antibodies targeting the N and C termini of the protein. Keeping in mind the normal turnover of ECM, class A biomarkers seem more appropriate for measurement of cytokines and growth factors that exist in their latent form in the circulation. Importantly, class A biomarkers can be encapsulated by macroglobulins, chaperones, or other transport or encapsulating proteins, which can mask the epitope of interest [32].

Class B biomarkers (Fig. 31.2B) of the ECM are probably used more often to detect protein fragments in the circulation than class A. However, several epitopes per protein exist. Class B can detect neoepitopes generated as part of the ECM turnover, modifications of the ECM proteins generated by pathological processes, or a combination of the two. The different subclasses of biomarkers can be detected by different proteomic approaches such as mass spectrometry and ELISA. ELISAs can be in both competitive and sandwich formats, because only one antibody is needed to detect a specific peptide sequence with a PTM on the antigen. The biomarkers targeting class B epitopes are often

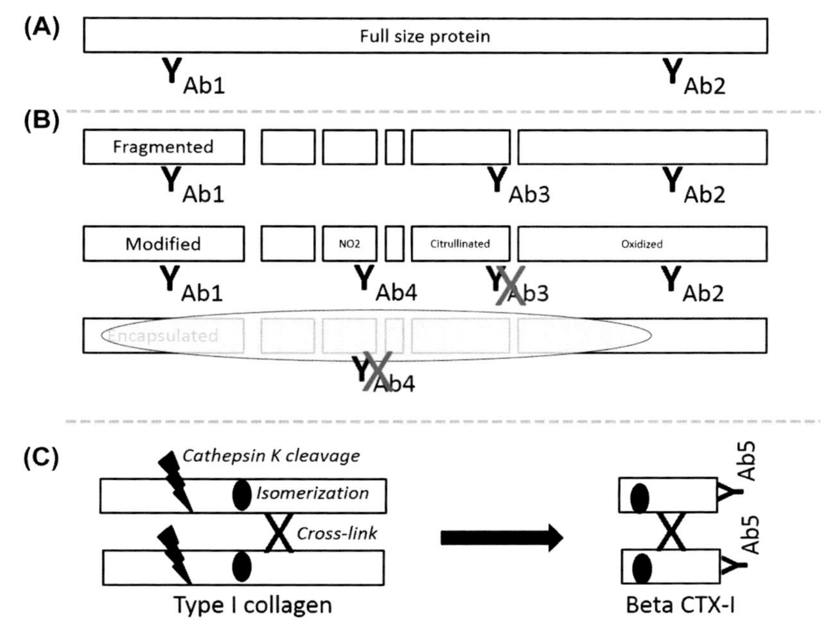

FIGURE 31.2 There are different types of extracellular matrix (ECM) biomarkers. (A) Measurement of full-size proteins often requires antibodies (Abs) with a broad range, eg, Ab1 targeting the N-terminal domain and Ab2 targeting the C-terminal domain. (B) Biomarkers of the ECM often target fragments of the molecule (Ab3). These fragments can be modified by different processes, such as oxidation, which then can be targeted as specific biomarkers (Ab4). A posttranslational modification process may destroy an epitope measuring the unmodified fragment (crossed Ab3). Also, some proteins are encapsulated by chaperones, masking the epitope of interest. (C) Isomerized type I collagen is the major collagen of the bone. It is a crosslinked triple helix that is degraded by the osteoclast enzyme cathepsin K.

greatly informative for prognosis, but they are seldom organ or tissue specific. An example is the biomarker VICM, a citrullinated and MMP-generated fragment of vimentin and a biomarker of macrophage activity [33,34].

Class C biomarkers (Fig. 31.2C) are tissue and potentially organ specific. Specificity may be achieved by a combination of different PTM processes that are pathospecific on proteins that are more abundant in the ECM in the tissue of interest. The identification of such biomarkers is cumbersome, and only a few examples of ECM-specific biomarkers are available. One example is β-CTX-I, which is described later in this chapter.

This separation of biomarkers into different classes is important because different technologies provide different levels of accuracy and precision. Numerous novel PTM processes on ECM proteins are discovered by proteomic approaches such as oxidation, citrullination, proteolytic fragmentation, glycosylation, and nitrosylation [35,36]. This suggests additional applied research is required to understand the exact analyte measured, because different PTMs may have different pathological meanings and measurement of different epitopes provides completely opposite information. Thus, different PTMs may, in part, explain why different versions of analytical methods thought to be measuring the same protein often lead to divergent results. A well-characterized example of this issue is measurement of type I collagen and its degradation fragments in postmenopausal OP [2,37]. Measurement of total type I collagen in serum would suggest that postmenopausal women have more type I collagen than age-matched individuals [37]. However, whether the increased levels are the consequence of degradation, or reduced formation of new type I collagen, is not revealed by a class A biomarker. Propeptides of proteins are released during protein synthesis and can be measured as surrogates for tissue formation. The N-terminal propeptide of type I collagen (PINP) is released as the protein is incorporated into the ECM; consequently, PINP in the circulation is used to measure both bone formation and fibrogenesis [38]. Meanwhile, the degradation of type I collagen is mediated by several proteinases, which yields several different degradation products of type I collagen. CTX-I is widely used in ELISA for the detection of cathepsin K–mediated destruction of type I collagen. Because the combination of type I collagen and cathepsin K is mainly observed in bone, CTX-I is considered a unique marker of bone resorption and is widely used as such [39]. Furthermore, at least two different versions of CTX-I exist, an α and a β version. Both measure the same cathepsin K–generated cleavages, but the β version measures a β-isomerized (oxidized) form of the fragment, whereas the α version measures the nonisomerized form. This can be translated into β-CTX-I, reflecting resorption of old bone as seen in OP and into α-CTX-I, reflecting the resorption of young bone, as seen in subchondral bone remodeling in osteoarthritis (OA) and bone metastasis [40,41]. Another biomarker of type I collagen destruction is type I collagen–derived crosslinked carboxy-terminal telopeptide (ICTP), which is a triple-crosslinked carboxyl-terminal telopeptide of type I collagen generated by MMP and destroyed by cathepsin K [24]. ICTP is mainly generated in inflamed connective tissue. Last, the C1M is released from the helical domain of type I collagen, also by the action MMPs [22]. The different fragments of type I collagen may provide different information, such as the occurrence of opposite processes of bone formation and bone resorption, by the different measurements of the same molecule.

PTMs OF ECM PROTEINS

Identification of tissue-, protein-, and disease-specific PTMs is needed for their optimal application as biomarkers to characterize pathogenesis and monitor for a disease. Because only a few proteins can be considered tissue specific, the combination of a tissue- or disease-specific protein with related specific proteases or factors offers the possibility to increase the sensitivity and specificity of an assay analyzing specific combinations of PTMs. Well-designed combinations of PTMs that are disease specific may provide a new generation of biomarkers. In Table 31.1, different potential combinations of PTMs in a single assay are proposed to be targeted in different pathologies.

TABLE 31.1 Composition of Synovial Fluid Relative to Other Human Bodily Fluids [115–117]

Test	Synovial fluid	Plasma	Sweat	Saliva	Semen
pH	7.4	7.35–7.45	4–6.8	6.4[a]	7.19
Specific gravity	1.008–1.015	1.0278	1.001–1.008	1.007	1.028
Electrolytes/mEq L^{-1}					
Potassium	4.0	3.5–5.0	4.3–14.2	21	31.3
Sodium	136.1	135–147	0–104	14[a]	117
Calcium	2.3–4.7	8–11	0.2–6.0	3	12.4
Magnesium		1.5–2.5	0.03–4	0.6	11.5
Bicarbonate	19.3–30.6	95–105		6[a]	24
Chloride	107.1	22–28	34.3	17	42.8
Proteins/mg dL^{-1}					
Total	1.72 g dL^{-1}	6–8 g dL^{-1}	7.7	386[b]	4.5 g dL^{-1}
Albumin	55–70%	50–65%			
α-1-Globulin	6–8%	3–5%			
α-2-Globulin	5–7%	7–13%			
β-Globulin	8–10%	8–14%			
γ-Globulin	10–14%	12–22%			
Hyaluronic acid	0.3–0.4 g dL^{-1}				
Metabolites/mg dL^{-1}					
Amino acids			47.6	40	1.26 g dL^{-1}
Glucose	70–110	70–110	3.0	11	224[c]
Uric acid	2–8	2–8	26–122[d]	20	72
Lipids, total				25–500[e]	188

[a]*Increases under salivary stimulation.*
[b]*Primary α-amylase, with some lysosomes.*
[c]*Fructose caption.*
[d]*Not present in eccrine secretions.*
[e]*Cholesterol.*

PTMs are the result of modifications to the composition or structure of proteins that expose new epitopes (ie, biomarker targets) called neoepitopes. Important processes creating PTMs are proteolytic cleavage (fragmentation); hydroxylation of prolines, resulting in hydroxyprolines; cleavage of procollagen peptides; crosslinking both intramolecular and intermolecular collagen fibrils [42–46]; age-related changes, such as isomerization and glycosylation [47,48], resulting in isoaspartic acids and advanced glycation end products such as pentosidine [47–50]; and nitrosylation of tyrosines. The PTM processes affect the protein at different levels. Some processes are important for maintenance of the tissue, whereas others are be destructive. This important difference has been described in cartilage turnover [16,51], where aggrecanase-mediated degradation aggrecan in explant cultures, measured by release of ^{374}ARGV fragments, of cartilage was reversible. However, when the disease stage had reached MMP-mediated tissue destruction, measured by release of ^{342}FFGV, cartilage repair was no longer possible. This exemplified that different fragments of the same protein may reflect different pathobiological or physiological processes [16,52].

Endopeptidase Activity

Specific proteolytic activities are a prerequisite for a range of cellular functions and interactions within the ECM. These specific activities are tightly coordinated under physiological situations in which a detailed sequence of events produces the adequate proteolytic response to promote tissue turnover. Examples are the complex proteolytic activities needed for cell migration and function [53]. This ECM remodeling generates a range of tissue and pathospecific turnover products that may be used as molecular biochemical markers. The proteolytic action of endopeptidases generates specific cleavage fragments. Even though many components of the ECM, as well as enzymes responsible for remodeling, are present in different tissues, the combination of a specific peptidase and specific ECM protein may provide a unique combination that elucidates activity in a particular tissue or a specific disease mechanism. The action of peptidases on ECM components results in matrix degradation fragments (cleavage neoepitopes). Previous examples of endopeptidase activity biomarkers have been addressed in previous sections of this chapter.

Crosslinking

Crosslinking plays an important part in the ECM meshwork and thereby in tissue integrity. Crosslinking is especially important for the fibrillar collagens such as types I and III and for minor collagens as types IV–XIV. The fibrillar collagens are characterized by their undisturbed triple helix that is stacked head-to-tail in connective tissue such as bone and cartilage [19]. This constitutes the backbone of the tissue. It is therefore important for collagen to be able to crosslink with the neighboring collagen as well as other ECM components [54]. Crosslinking between different ECM components or between different protein chains can result from enzymatic and nonenzymatic pathways. Enzymatic crosslinking is often processed by lysyl oxidase. When the fibrils of collagen are aligned, aldehydes on lysines and hydroxylysine can react with corresponding residues forming aldol condensation products [55]. Nonenzymatic glycosylation of fibrillar collagen as well as minor collagens are also important crosslinking residues, which are increased in several pathological conditions such as diabetes and fibrotic diseases [45,47,56,57].

Nonenzymatic Glycosylation

Nonenzymatic glycosylation, also called the Maillard reaction, is a naturally occurring phenomenon and leads to PTM of proteins, nucleic acids, and lipids [58]. A common cause of nonenzymatic glycosylation is increased blood glucose levels; accordingly, most knowledge about nonenzymatic glycosylation arises from studies performed in diabetics [58]. Reducing glycosylated hemoglobin (HbA_{1c}) is the primary objective when treating diabetics [58]. Aging is also associated with accumulation of glycosylations, and recent studies highlighted that these glycosylations are associated with a range of diseases that typically afflict older populations, including OP, OA, Alzheimer, cardiovascular disease, chronic heart failure, liver fibrosis, and various nephropathies [59,60].

Nonenzymatic glycosylation is initiated by reducing sugars such as glucose, fructose, and pentose through reactions with the NH_2 side chain containing the amino acids arginine, lysine, and hydroxylysine. This leads to the formation of a Schiff base that then is converted into an Amadori product, also called an early glycosylation product [61]. HbA_{1c} is a commonly used example of an early glycosylation product, but because of the numerous publications about the relevance of HbA_{1c}, it will not be described in further detail [58].

The Amadori product is then degraded, resulting in an available NH_2 group on the protein and a class of highly reactive dicarbonyls such as glyoxal, methyl glyoxal, or 3-deoxyglucosone that are referred to as intermediate glycosylation products [58,61]. Methyl glyoxal has been indicated to be a very important intermediate glycosylation product, primarily because it is a major precursor for advanced glycation end products (AGEs) (see below), but also because of its intrinsic effects on pancreatic beta cells as well as other cells [62–64]. The most important role of the intermediate glycosylation products, which are reactive carbonyls, is their reaction with the NH_2 side chain–containing amino acids, leading to irreversible AGEs [58,61,65]. If oxidation occurs at the time of the reaction, a category of AGEs sometimes referred to as glycoxidation products, are formed. These include pentosidine and N^{ε}-carboxymethyllysine (N^{ε}-CML), which represent two classes of AGEs, namely, the crosslinking and noncrosslinking types, respectively. Pentosidine and N^{ε}-CML are the two best-described AGEs [59]. AGEs consist of long-lived proteins and often cause damage by forming irreversible crosslinks between individual proteins, ie, between the fibers in the triple helix of type I collagen [66]. In addition to crosslinking, the AGEs also activate specific receptors that presently are categorized into either inflammation or clearance receptors. Several AGEs have been described in the literature [61].

A more recent area of research for pentosidine is bone biology. Because pentosidine is a crosslinking AGE, it introduces molecular bonds between two long-lived proteins [67] such as collagen. In bone, collagen type I is, by far, the most abundant molecule. It has a very long half-life, depending on bone remodeling rates [66], thereby rendering it a prime candidate for nonenzymatic glycosylation. Accordingly, pentosidine crosslinks are numerous in aged bone samples [68]. Furthermore, accumulation of pentosidine in collagen type I is associated with a deterioration in the functionality of the molecules [69]. The crosslinks formed in type I collagen are associated with increased stiffness of the collagen helices; and in bone, increased stiffness is associated with increased probability of fracture [66,68,70].

Compared with age-matched healthy subjects, OP patients have been found to have elevated serum levels of pentosidine, indicating that pentosidine is a serum marker of OP [71]. A study performed in elderly Japanese women indicated that urinary pentosidine levels were predictive of fracture risk [69]. In contrast, the OFELY study indicated that urinary pentosidine levels do not indicate increased fracture probability in women [72]; therefore, the utility of pentosidine as a biochemical marker for OP is not clear and requires further research.

Type 1 and type 2 diabetics have poor bone quality and increased fracture rates [73]. Interestingly, pentosidine has been speculated to be a causal link between diabetes and bone fractures [73,74]. Low serum levels of AGE receptor are also associated with increased probability of fracture in type 2 diabetics [75]. Because studies have shown that antibodies toward AGEs can be developed in mice [76–78], we believe these markers, combined with the correct enzyme-generated neoepitope, will likely provide highly accurate assays for monitoring different diseases in which glucose levels and resulting modifications are involved.

Oxidations and Hydroxylations

Oxidative damage to proteins is often caused by the action of free radicals, reactive oxygen species (ROS) such as nitric oxide, or hydrogen peroxide. Oxidation has been implicated in several pathological and healthy tissue turnover processes. Although many amino acids can be attacked by ROS, some seem more likely to undergo oxidation than others. For example, lysine and proline are readily oxidized to aldehydes and sulfoxidation of methionine and nitrosylation of tyrosines also occurs [57]. Under normal conditions, these ROSs are strictly regulated by antioxidants such as peroxidases and dismutases, among others [79]. However, under pathological conditions, oxidation may play an important role in tissue destruction.

Structural proteins such as collagens are primary targets for superoxide dismutase-catalyzed nitration by peroxynitrite. When collagen chains are disassembled due to structural changes in the tissue, eg, due to pathological change, the tyrosines are exposed to their surroundings and thereby become susceptible to nitration [80]. The reaction of nitrogen to superoxide-producing peroxinitrite leads to the nitration of tyrosine residues in proteins, resulting in the formation of the posttranslationally modified nitrotyrosine (Fig. 31.2A). Nitrotyrosines are negatively charged, and this can lead to further disruption of the collagen [80]. Beckman et al. have shown that extensive nitration takes place around foamy macrophages in human atherosclerotic lesions and early fatty streaks, and also in the foci of myocytes in the vascular smooth muscle [81]. Thus, tyrosine nitration may be an opportune marker of reactive nitrooxidants being produced by pathological processes. Although peroxynitrite is not the only nitroreactive element, it is the most likely source in vivo [80]. Other reactive species such as nitrogen dioxide or nitrate can form nitrotyrosine in simple solutions. However, the amounts of nitrogen dioxide or nitrite that are present in vivo are far lower than necessary to cause significant nitration in vitro.

Biochemical marker assays against nitrotyrosine in type III collagen have been developed. In some cases of arthritis, levels of nitrosylated type III collagen are elevated, and because type III collagen is turned over at a relatively high rate, the level of nitrosylated type III collagen can be measured in the circulation [82]. Richardot et al. showed that this neoepitope was elevated in serum of patients with arthritis and that the level significantly correlated with serum C-reactive protein (CRP) [82]. Similarly, elevated levels of nitrosylated type II collagen have been detected in RA and OA patients and were also in these studies correlated with CRP [83,84]. Circulating nitrosylate fragments of type II collagen have been shown to be significantly associated with disease severity in OA [83,84].

Isomerization: Age of ECM proteins

PTMs accumulate with age [36]. The rate at which transformations of proteins occur and the fragments eventually build up is controlled by the level of tissue turnover. In tissues with well-balanced turnover, PTMs normally contribute to crucial cell signaling, protection of peptides against proteolysis, and strength of ECM—all components of matrix integrity [36]. Nevertheless, it has become well acknowledged that aging of proteins has an important impact on cell interactions at the molecular level [85] and that PTMs can be involved in various disease pathologies [86]. As an example, autoimmune responses to a specific PTM (as described later) are known to be associated with the hallmarks of protein aging [85,87,88]. Peptides that contain amino acid isomerizations are often resistant to proteolysis [35], and this feature affects the procession of antigens for presentation on the major histocompatibility complex II involved in the immune-response signaling for the production of T cells and antibodies [36]. In preclinical studies, it has been shown that various known autoantigens contain sites prone to deamidation and that isomerization is involved in type I diabetes, RA, systemic lupus erythematosus, and experimental autoimmune encephalomyelitis [35,89]. Thus, isomerization plays an important role in some common diseases.

Proteins containing aspartate (D), asparagine (N), glutamate (E), or glutamine (Q) residue linked to a low-molecular-weight amino acid, such as glycine (G), can undergo a spontaneous nonenzymatic isomerization (Fig. 31.2C) [36]. This isomerization introduces a kink in the conformation of the molecule, because the peptide backbone is redirected from the α-carboxyl group in the native newly synthesized form to the side chain β-carboxyl [48,90].

Citrullination and Deiminations

Citrullination or deimination of proteins is the reaction for converting the amino acid arginine into the amino acid citrulline. This reaction is catalyzed by enzymes called peptidylarginine deiminases [91,92]. The conversion of arginine into citrulline can have important consequences for the structure and function of proteins, because arginine is positively charged at a neutral pH, whereas citrulline is uncharged. This increases the hydrophobicity of the protein, leading to changes in protein folding. Proteins that normally contain citrulline residues include myosin basic protein, fillagrin, and several histone proteins, whereas other proteins such as fibrinogen, fibrin, and vimentin can be citrullinated during cell death and tissue inflammation [93].

Citrullinated protein antigens are the dominant targets of the autoimmune response in RA [94]. The presence of antibodies to citrullinated proteins is detected by anticyclic citrullinated peptide (CCP) assays. These constitute a powerful diagnostic tool for RA, surpassing in both sensitivity and specificity the rheumatoid factor [95]. A key characteristic of anti-CCP antibodies that adds to their overall value is their increased prognostic capacity. Evidence highlights the increased sensitivity and specificity of anti-CCP2 assays [91,92,96]. The test for the presence of anticitrullinated protein antibodies is highly specific (88–96%) and is about as sensitive as the rheumatoid factor (70–78%) in diagnosing RA. They are detectable even before the onset of clinical symptoms.

Numerous ECM remodeling–related diseases are considered as autoimmune disorders, eg, cardiovascular disease [97]. In RA, inflammation is one of the main clinical manifestations. The great success that citrullinated proteins have had in RA diagnosis and prognosis could potentially be replicated in other diseases with similar clinical manifestations. In fibrotic, cancerous, or atherosclerotic tissue, citrullinated proteins that act as a connective link triggering alone or in conjunction with other PTMs an immune response, could be potential new targets for biomarker development. The potential benefits could include the identification of new ECM remodeling biomarkers with increased tissue specificity when used in conjunction with ECM degradation fragments.

DIFFERENT MATRICES MEAN DIFFERENT THINGS

The measurement of the same biomarker in different matrices can generate very different results; thus, it is important to understand which material is being assessed.

Serum Versus Plasma

Different blood-based matrices such as serum and plasma may be chosen for analytical samples. Despite serum being the most common matrix used, it may not always be suitable, particularly if the analyst is of hemostasis origin. In this case, plasma is the preferred matrix. Serum specimens are obtained by allowing the whole blood to clot, during which time various amounts of proteins are removed into the fibrin clot; so, the remaining, nonclotted sample may not contain the analyst. Unlike serum, however, plasma contains fibrinogen together with other procoagulant proteins that will result in coagulation unless subjected to an anticoagulant. The protein concentration of serum is therefore less than that of plasma [98,99]; thus, considerations should always be given to the analyst and the choice of matrices. Plasma constitutes 55% of the total blood volume and serves as a solvent for electrolytes, nutrients, and proteins important for blood clotting and immunity. Plasma comprises 92% water. It is obtained by separating the liquid portion of the blood from the cells by the use of an anticoagulant.

Numerous anticoagulants and variations of these may be employed for obtaining plasma samples and the most well-known are heparinate, citrate and ethylenediamine tetraacetic acid (EDTA). These anticoagulants inhibit blood coagulation by impeding different phases of the coagulation cascade. Considerations to the choice of anticoagulant should be given as any one anti-coagulant may influence the analyte outcome. This is seen with the variation in plasma concentrations of matrix metalloproteinases and their tissue inhibitors upon the use of different anticoagulants [100,101]. The choice of anticoagulant may be even more important if the analyst is of coagulation activation origin. The concentration of these analysts may vary depending on the used anticoagulant [102], because anticoagulants differ in their ability to reduce clotting. Thus, a cross-matrix comparison is always desirable to allow for precision and accuracy of the analyst and to avoid selection of an inappropriate anticoagulant.

A general drawback with plasma is that the chemical composition upon long-term storage may change and fibrin clot formation may occur. In addition, systematic errors such as a clotted sample due to improper sampling may lead to inaccurate analyst determination. Many hemostasis methods display high sensitivity to small preanalytical variation, as has been seen with repeated sampling from the same patients [103]. In fact, preanalytical irregularities such as incoherent coagulation may be associated with inconsistent laboratory results, leading to erroneous clinical interpretation. It is estimated that 32–68% of all laboratory errors occur during the preanalytical phase [104]. This is one reason to why serum is in general preferred. However, the choice of matrices depends on the analyst in question.

Urine

There is evidence that the levels of neoepitope markers measured in urine and in serum or plasma can be radically different in animal models of kidney disease [105] and in patients with immunoglobulin A nephropathy [106]. This may reflect actual pathological processes. Urine is the natural matrix of the kidney; therefore, information obtained from proteins excreted or secreted from the affected kidneys into urine could describe the disease status, whereas proteins released into the circulation might reflect more systemic events such as inflammation. Moreover, different organs can contribute to the pool of peptides and proteins that are found in circulation [107].

Serum and plasma have a wide dynamic range of protein contents compared with urine. Albumin accounts for approximately 50% of the total mass of proteins in the serum, and the most abundant 22 proteins are estimated to account for 99% of the serum protein mass [107]. Urine, in contrast, has a much lower content of proteins, at least in healthy individuals with an intact filtration barrier.

These disparities in protein content might also be explained by the fact that because urine is the product of blood filtration, peptides might go through further processing during filtration and during their permanence in the urine, and this can be reflected in the measurements of samples. This is particularly true in the case of neoepitope markers, which are detected in urine only if the neoepitope is not further degraded. At the same time new neoepitopes might arise after the processing of peptides and proteins in the kidney, and the marker can therefore be measured only after these events.

It is also necessary to take into account that metabolites filtered in the urine can influence the assays used to measure the biomarkers. An example of this is the alteration in levels of urine creatinine measured by the creatininase-based reaction assay (but not by the Jaffé-based reaction assay) in urine of patients with alkaptonuria. Urine from these patients contains high levels of homogentisic acid [108]. This had been previously observed in routine clinical chemistry assays in urine [109]. For these reasons, it is important to, if possible, measure the biomarkers in both serum and urine and not to automatically translate the results obtained in one matrix to the other.

Synovial Fluid

Synovial fluid (SF) is the viscous liquid in the synovial cavity and is secreted by the synovial membrane. Its function is to reduce friction between the articular cartilages of the synovial joint during movement. It is a dialysate from plasma to which components produced locally by joint tissue is added. The composition of the SF changes during an individual's lifetime due to changes in mechanical loading [110,111] and with joint degradative diseases. The main components of SF are water, hyaluronic acid (HA; roughly 3–4 mg/mL), D-glucuronic acid, D-*N*-acetylglucosamine, growth hormone prolactin, and lubricin [112]. Because SF is the fluid covering the tissues of the joint, it is closer to the cartilage and synovial membrane than blood. In addition, cartilage is an avascularized tissue; thus, the fragments of cartilage need to pass through the SF to reach blood. This means that measuring fragments of cartilage and synovial membrane in SF is a more direct measure of the tissue turnover than an assessment in blood. In theory, SF is preferred over blood for assessment of joint turnover, especially cartilage and the synovial membrane. However, the very complex and viscous SF matrix can easily hide peptides and crystals, which complicates their assessment. There is no gold standard pretreatment of the SF before analysis; thus, several methods for analyzing protein and crystal content in SF are used. Even though retrieval of SF is fairly easy, it is not a common procedure in standard care of joint diseases.

Hyaluronidase pretreatment is the most common pretreatment. It cleaves the HA in SF into smaller fragments, giving a less viscous liquid that is easier to work with in experiments. HA is a long polymer, which very substantially causes the viscous properties of SF. As a consequence of loading, associations of polymer chains of HA occur and rheopexic properties of SF are manifested [113]. The production of HA reaches its peak during adolescence and thereafter declines with age. Ultrasonication is another pretreatment method of SF [114]. Another method is to purify samples before using scanning electron microscope or atomic force microscope for assessment of crystals [115]. Centrifugation after hyaluronidase pretreatment is another method that has been used to assess crystals [116], and pretreatments with papain and sodium hypochlorite have been used to assess crystals [117,118]. Last, a method where samples were subjected to deglycosylation with chondroitinase ABC, and keratinase has been used for assessment of aggrecan peptides [119]. All of these methods have been individually optimized for the specific peptide or crystal to be assessed, which illustrates the complexity of analyzing SF.

In summary, SF is a complex, viscous fluid that offers potential for the assessment of cartilage and synovial membrane turnover. However, a standard method for pretreatment of the fluid is needed.

Bronchoalveolar Lavage and Sputum

Bronchoalveolar lavage (BAL) is a method that enables the investigation of cellular and acellular components of the lower respiratory tract by administration of saline. BAL retrieves the secretions that coat the surface of bronchial and alveolar epithelium, diluted by the instilled saline. The technique allows direct insight into processes involved in injury of the lung tissue including inflammation, tissue degradation, and changes in the permeability of the blood–air barrier. BAL is easily performed, relatively reproducible, well tolerated, and can even be done in acutely ill patients [120]. Compared with a lung biopsy, it is less invasive, it may be done repeatedly, and it enables the sampling of a greater lung area. In healthy individuals, a BAL sample mainly consists of alveolar macrophages. Changes in the cell profile may support the diagnosis of a specific interstitial lung disease [120–123]. Proteins in the BAL fluid may be produced locally in the lung or originate from plasma, presumably derived from diffusion across the blood–air barrier. Permeability may increase with disease due to inflammatory damage, resulting in elevated plasma protein levels in BAL as a consequence of increased protein leakage from the bloodstream to the lung tissue [124,125]. Both cellular profile and alterations in protein expression have been demonstrated to relate to various lung diseases including idiopathic pulmonary fibrosis, sarcoidosis, asthma, and cystic fibrosis [126–129]. Although BAL is a fairly simple matrix with a basis of saline, the amount collected in each sample may vary greatly due to the choice of lung site, quantity of instilled fluid, and type of aspiration. Thus, results should be expressed with a reference value such as milliliters of BAL recovered or per microgram of total protein to account for differences in recovery [125]. The considerable variability in the techniques used for the procedure as well as sample handling and processing may lead to great variations from lab to lab [123]. In 2012, the American Thoracic Society published a guideline on the clinical utility of BAL in interstitial lung diseases. it included a standardization, handling, processing, and interpretation of BAL, which hopefully will result in more uniformed analyses in future [120].

An alternative to BAL for investigating the airways directly is the collection of induced-sputum samples. A standardized procedure for the effective collection of a sputum sample using inhalation of nebulized saline has reduced the variability of the method [130,131]. Sputum collection is relatively noninvasive, safe, and reproducible, and it can be collected repeatedly even during severe disease or exacerbations. Evaluation of airway inflammation in sputum has been widely used in obstructive lung diseases such as asthma and COPD [132,133]. Sputum processing is laborious and includes the addition of dithiothreitol (DTT) to dissociate the disulfide bonds of the mucus and thus homogenize the sample. As with BAL, variations in the sputum processing technique may cause variable results [134]. Studies have indicated that sputum measurements may be more reproducible than BAL measurements [132,135–137]. A study comparing BAL, induced sputum, and blood measurements of inflammatory cells and proteins in mild asthmatics resulted in differences presumably related to measurements of different compartments of the airways, which did not necessarily correlate [138]. The authors propose that BAL samples should be used for investigations of the peripheral airways and sputum samples for the larger airways. Conversely, blood measurements reflect systemic inflammation that does not seem to fully reflect the disease-associated inflammatory processes within the airways [137]. Sputum sampling have an advantage over BAL in that it shows higher cell recovery and a stronger signal of acellular components [138]. However, sputum samples may be contaminated by saliva that may influence measurements of both cells and proteins [139,140]. Sample analysis using immunoassays may also be affected by the content of DTT as this may disrupt the disulfide bonds of the antibody used.

APPLYING STRUCTURAL BIOMARKERS AS DIAGNOSTIC OR MONITORING TOOLS

Bone and skeletal disease were probably the first to use biomarkers measuring different biomarkers of type I collagen. Bone biomarkers have been used to test cathepsin K inhibitors and other antiresorptives for the treatment of OP, such as ONO-5334 [140a], denosumab [141] and odanacatib [142]. Cathepsin K is an osteoclast collagenase that mediates bone resorption measurable by CTX-I or N-terminal telopeptide of type I collagen (NTX-I) in either serum or urine of patients. For example, ONO-5334 was tested in 197 postmenopausal women with OP or osteopenia with one fragility fracture. After 24 months, ONO-5334 was shown to be associated with increased bone mineral density for lumbar spine, total hip, and femoral neck ($P<.001$), whereas the levels of both serum and urine CTX-I, and urine NTX-I as early as 2 weeks after initiation of treatment. In patients treated with ONO-5334, the bone-resorption markers urinary (u)NTX and serum and uCTX-I were significantly suppressed throughout 24 months of treatment and to a similar extent as was seen in patients receiving alendronate [143,144]. Another type I collagen biomarker, initial carboxyl-terminal telopeptide, levels were increased in patients receiving ONO-5334. Cathepsin K inhibition with ONO-5334 resulted in decreases in most resorption markers over 2 years, but did not decrease most bone formation markers. This was associated with an increase in radiographic bone mass density. The effect on biomarkers was rapidly reversible on treatment cessation. This is a recent example of how biomarkers of the same protein provide different information and can be used as early indicators of treatment efficacy.

Another example of a type I collagen serological biomarker is C1M [22], which was tested in the LITHE study. This was a 1-year phase III, double-blind, placebo-controlled parallel group study of tocilizumab (TCZ) 4 or 8 mg/kg every 4 weeks, in RA patients on stable doses of methotrexate (MTX) [145]. At baseline, C1M was significantly ($P<.0001$) correlated to CRP, to a visual analogue scale pain ($P<.0001$), disease activity score (DAS28-ESR) ($P<.0001$), joint space narrowing (JSN) ($P<.01$), and modified total Sharp score (mTSS) ($P<.001$). These correlations indicate that C1M is marker of disease activity [146]. In addition, baseline C1M was significantly correlated with change in JSN or mTSS from baseline to week 24 ($P<.0001$) and to week 52 ($P<.0001$). C1M levels were dose dependently reduced in the TCZ+MTX group. Thus, baseline C1M levels correlated with worsening joint structure over 1 year. Serum C1M levels may enable identification of those RA patients who are in most need of aggressive treatment.

Type II collagen is almost solely produced by the cartilage cells, chondrocytes. Great efforts have been made to develop specific biomarkers targeting different epitopes of type II collagen and thereby measures of type II collagen turnover [147,148]. Urinary CTX-II is one of most tested biomarkers and has been used in several preclinical and clinical studies. Serum PIIANP and urinary CTX-II were measured in the CIMESTRA study that recruited early RA patients to investigate the relationship between markers of collagen II formation and degradation with disease activity measures and radiographic outcomes at a 4-year follow-up [149]. PIIANP was low at diagnosis and 4 years on ($P<.001$), irrespective of treatment and disease activity. CTX-II at baseline was increased ($P<.001$) and correlated positively with disease activity and radiographic progression, but not with anti-CCP. These results suggest that type II collagen formation and degradation are unbalanced when RA is diagnosed. CTX-II has also been used as a pharmacokinetic and pharmacodynamics marker in different settings. The bioavailability and pharmacodynamics of synthetic salmon calcitonin and recombinant salmon calcitonin were tested in healthy postmenopausal women. Measurement of bone resorption by using the CTX-I and CTX-II biomarkers displayed comparable responses, with areas under the curve of relative change of serum CTX-I of −250% × hours and relative change in urine CTX-II of −180% × hours during the 4-h observation period. The biomarker data led investigators to conclude that oral synthetic and recombinant calcitonin displayed comparable pharmacodynamic and kinetic properties [150]. Several other studies have used CTX-II to monitor and compare treatment effect in patients with joint diseases [151,152].

Urinary type II collagen neoepitope (TIINE) is another biomarker of type II collagen degradation. It has been used to investigate the mechanism underlying the slowing of JSN in patients with OA. Otterness et al. tested the effect of doxycycline in knee OA [153–155]. The mean TIINE concentration for doxycycline-treated patients was higher than that for the placebo group. Hellio Le Graverand et al. examined whether uTIINE could distinguish subjects with progressive radiographic and/or symptomatic knee OA from those with stable disease [156]. Baseline uTIINE levels were unrelated to JSN in the placebo group. However, among subjects in the active treatment arm, a 1 SD increment in baseline uTIINE (68 ng/mM Cr) was associated with a marginally significant, two-fold increase in the odds of progression of JSN (odds ratio 2.04). The within-subject mean of uTIINE values at baseline, 6 months, and 12 months was associated with concurrent JSN measured at 16 weeks. Although baseline uTIINE was not a consistent predictor of JSN in subjects with knee OA, serial measurements of uTIINE reflect concurrent JSN.

Other biomarkers of type II collagen turnover have been used. Anti-TNF therapies provide symptomatic benefit for patients with spondyloarthropathy (SpA). The effect of etanercept on biomarkers of type II collagen synthesis and degradation in patients with SpA followed for 2 years was investigated by Briot et al. [157,158]. Cartilage degradation was investigated by measuring serum levels of the type II collagen fragments and C2C, whereas the C-terminal propeptide of type II collagen (PIICP) was used as a marker of type II collagen synthesis. Over 2 years, there was a significant decrease of serum C2C ($P<.01$). Compared with baseline, the decrease in serum C2C was significant at month 12 (−12.1%; $P<.01$). Conversely, PIICP increased significantly by 17% ($P<.01$) from baseline to 24 months. These data suggest that etanercept may have beneficial effects on cartilage metabolism in patients with SpA. These data have been supported by several follow-up studies measuring the biomarkers C2M and ProC2 in SpA patients [31].

Hu et al. [159] compared apparent diffusion coefficient (ADC) values on diffusion-weighted imaging from hepatic fibrosis patients with those from healthy controls. The purpose was to identify correlation between ADC values with serum indices of liver fibrosis. HA, laminin (LN), type III procollagen (PCIII), and collagen type IV (IV-C) were measured in 54 hepatic fibrosis patients and 23 healthy controls. With progressive liver fibrosis, PCIII levels were increased ($P<.01$). There were negative correlations between ADC and LN, PCIII, HA, and IV-C.

During hepatic fibrosis progression, the quantity and quality of the hepatic ECM changes with an up to five-fold increase in total collagen content along with a change in the collagen profile resulting in twice the amount of type I collagen compared with type III collagen [160,161]. The increase in collagen deposition is subsequently followed by a shift in matrix composition from the low-density basement membrane–like matrix to an interstitial matrix containing fibril-forming collagens [162].

Regardless of etiology, cirrhosis is the end stage of progressive fibrogenesis. However, the developmental pattern of fibrosis depends on the underlying etiology [163]. Hepatic stellate cells (HSCs) have long been considered the main fibrogenic cell type, possibly due to their ability to become isolated from human and rodent liver tissue. Most of the knowledge related to hepatic fibrosis has been based on in vitro activation of HSCs. Several fibroblast-like cell types contributing to fibrosis development have been identified, including septal and interface myofibroblasts and smooth muscle cells [164–167].

Nielsen et al. assessed two markers (Pro-C3 and C3M) of formation and degradation of type III collagen in baseline serum and stratified according to the baseline Ishak scores. They found the two markers had similar diagnostic performances as aspartate transaminase, alanine transaminase, and FibroTest. Interestingly, when stratifying the baseline marker levels to the changes in Ishak scores after 52 weeks, Pro-C3 was the only marker that significantly differentiated progressors from stable patients. Patients with high levels of Pro-C3 at baseline had more than four times higher odds of being progressors of fibrosis than those with low baseline levels [168].

Another example of the use of biomarkers is in the diagnosis of hernias, where a disturbed metabolism in the ECM contributes to an abnormal formation of the abdominal wall. A recent study investigated the level of ECM biomarkers in patients with different types of hernia: Primary unilateral inguinal hernia, multiple hernias defined as three or more hernias, and incisional hernia. Biomarkers for synthesis of interstitial matrix (PINP, Pro-C3, Pro-C5) and basement membrane (P4NP) as well as corresponding degradation (C1M, C3M, C5M, and C4M) were measured in serum. In inguinal hernia patients, the turnover of the interstitial matrix collagens types III ($P<.05$) and V ($P<.001$) was decreased compared with controls, whereas the turnover of the basement membrane IV-C was increased ($P<.001$). In incisional hernia patients, the turnover of type V collagen was decreased ($P<.05$) and the turnover of type IV collagen was increased compared with the hernia-free controls ($P<.001$). Hernia patients demonstrated systemically altered collagen metabolism. The serological turnover profile of IV-Cs may predict the presence of inguinal and incisional hernia. The results indicated that regulation of IV-C turnover may be crucial for hernia development and may provide novel insight into the pathogenesis of hernia.

These examples highlight that ECM biomarkers using the generation of PTMs may provide novel insights into the pathology of disease as well as act as diagnostic or prognostic biomarkers.

BIPED CLASSIFICATION IN OA

The BIPED [4,5] categorization emphasizes the different biomarker roles: burden of disease, investigative, prognostic, efficacy of intervention, and diagnostic. Here, attention is drawn to the most important parameters for clinical trials, by using OA as an example First, to select the study population, diagnostic and prognostic markers are needed. A diagnostic (or alternatively a burden of disease) marker must ensure that the population is at the desired or well-defined stage of OA at baseline. A prognostic marker will likely report that the population has a high risk of progressing during the study period if they remain untreated. Finally, before final phase III clinical studies, an efficacy of intervention marker is needed to evaluate efficacy (Fig. 31.3).

The development of diagnostic, burden of disease, or prognostic biomarkers requires well-defined clinical trials, often cross-sectional case–control studies. It is therefore important that several trivial data are collected from the subjects included, such as sex, age, disease severity, and treatment history, to define a cohort within the general population and thereby narrow the field of interest (Fig. 31.3). The purpose of the diagnostic marker in this respect would be to distinguish between different states (eg, diseased vs nondiseased, treated vs nontreated, mild vs severe disease) and isolate patients from the general population. In contrast, burden of disease markers should be able to distinguish between severities or extent of disease within the group that were "positive" for the diagnostic marker. A prognostic marker can ideally predict the outcome from a given baseline situation of a patient by calculating the relative risk and odds ratio. It is consequently important that the biomarker or biomarkers chosen for the study are limited to those considered to be a gold standard. In OA, this

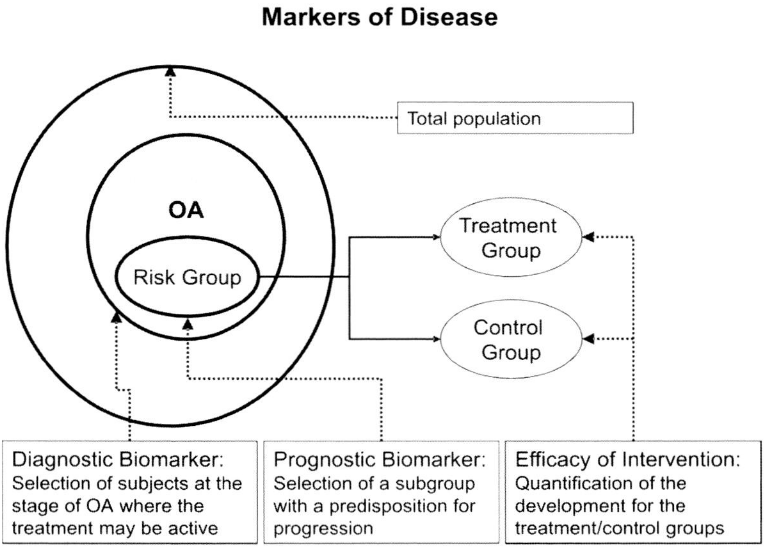

FIGURE 31.3 Three major classifications of the BIPED classification [4].

gold standard is often radiographic (eg, Kellgren–Lawrence scoring), which is not optimal because its accuracy is very much dependent on the observer. Both diagnostic and prognostic biomarkers should have high sensitivity and specificity for differentiation of diseased and nondiseased subjects. This is where many of the existing biomarkers tend to fail: many have shown that two cohorts have significantly different mean levels of a given biomarker, but failed to identify a relatively high percentage of individual patients as belonging to either cohort. In reality, a panel of markers is needed for predicting disease progression for an individual patient.

When a risk group has been defined by, eg, diagnostic and prognostic markers, a clinical trial can be taken a step further to test the effect of a potential treatment in randomized controlled trials. One such example was to investigate whether the progression of OA could be halted by treatment with bisphosphonates. The correlation between the outcome and the marker data at baseline will define whether the marker is an efficacy of intervention marker (Fig. 31.3).

In clinical trials, measuring biochemical markers in body fluids has typically several advantages over more conventional assessment tools such as radiography: (1) the collection of body fluid, such as blood or urine, samples is noninvasive, and can therefore be repeated over long-running studies; (2) test results can be achieved quickly; (3) the development of the disease can be followed continuously; and (4) on the basis of statistical power calculations, a relatively small sample size is needed to complete biomarker analysis.

In addition, validated and qualified diagnostic and prognostic biomarkers are useful tools in new drug development both in preclinical and clinical phases. Ideally, the markers should enable more or less effortless identification of subjects with specific disease characteristics and predict the effect of the candidate drug, its suspected side effects, or both. Furthermore, biomarkers identified for one disease might be applied to another disease, if it is suspected that the other disease also affects the release or representation of the same biomarker.

CONCLUSION AND PERSPECTIVES

In this book, we have highlighted the possibilities that are emerging in clinical chemistry, by the combination of multiple disease-specific neoepitopes in third-generation biochemical marker assays. This approach has already been applied to some disease areas such as bone and fibrosis, and it may be advantageous in yet other disease areas. By incorporating the most optimal biochemical markers in all aspects of drug discovery and development, with the promise of translational science, novel treatment opportunities may be identified, and drug research efforts may be stimulated by the ease of early detection of both efficacy and safety concerns.

Biochemical markers based on the advanced disease–tissue neoepitope approach may be one important tool to be used in combination with others. This approach may bring added recognition of the value of in vitro diagnosis and enable the integration of better markers into more aspects of research and drug development.

REFERENCES

[1] Karsdal MA, Qvist P, Christiansen C, Tanko LB. Optimising antiresorptive therapies in postmenopausal women: why do we need to give due consideration to the degree of suppression? Drugs 2006;66(15):1909–18.

[2] Schaller S, Henriksen K, Hoegh-Andersen P, Sondergaard BC, Sumer EU, Tanko LB, et al. In vitro, ex vivo, and in vivo methodological approaches for studying therapeutic targets of osteoporosis and degenerative joint diseases: how biomarkers can assist? Assay Drug Dev Technol 2005;3:553–80.

[3] Karsdal MA, Henriksen K, Leeming DJ, Mitchell P, Duffin K, Barascuk N, et al. Biochemical markers and the FDA critical path: how biomarkers may contribute to the understanding of pathophysiology and provide unique and necessary tools for drug development. Biomarkers 2009;14(3):181–202.

[4] Bauer DC, Hunter DJ, Abramson SB, Attur M, Corr M, Felson D, et al. Classification of osteoarthritis biomarkers: a proposed approach. Osteoarthritis and Cartilage 2006;14(8):723–7.

[5] Veidal SS, Bay-Jensen AC, Tougas G, Karsdal MA, Vainer B. Serum markers of liver fibrosis: combining the BIPED classification and the neo-epitope approach in the development of new biomarkers. Dis Markers 2010;28(1):15–28.

[6] Karsdal MA, Henriksen K, Leeming DJ, Woodworth T, Vassiliadis E, Bay-Jensen AC. Novel combinations of post-translational modification (PTM) neo-epitopes provide tissue-specific biochemical markers–are they the cause or the consequence of the disease? Clin Biochem 2010;43(10–11):793–804.

[7] Kraus VB, Blanco FJ, Englund M, Henrotin Y, Lohmander LS, Losina E, et al. OARSI clinical trials recommendations: soluble biomarker assessments in clinical trials in osteoarthritis. Osteoarthritis Cartilage 2015;23(5):686–97.
[8] Karsdal MA, Bay-Jensen AC, Leeming DJ, Henriksen K, Christiansen C. Quantification of "end products" of tissue destruction in inflammation may reflect convergence of cytokine and signaling pathways – implications for modern clinical chemistry. Biomarkers 2013;18(5):375–8.
[9] Bonilla-Hernan MG, Miranda-Carus ME, Martin-Mola E. New drugs beyond biologics in rheumatoid arthritis: the kinase inhibitors. Rheumatology (Oxford) 2011;50(9):1542–50.
[10] Madsen SH, Sondergaard BC, Bay-Jensen AC, Karsdal MA. Cartilage formation measured by a novel PIINP assay suggests that IGF-I does not stimulate but maintains cartilage formation ex vivo. Scand J Rheumatol 2009;38(3):222–6.
[11] Mocsai A, Ruland J, Tybulewicz VL. The SYK tyrosine kinase: a crucial player in diverse biological functions. Nat Rev Immunol 2010;10(6):387–402.
[12] Zou W, Kitaura H, Reeve J, Long F, Tybulewicz VL, Shattil SJ, et al. Syk, c-Src, the {alpha} v{beta}3 integrin, and ITAM immunoreceptors, in concert, regulate osteoclastic bone resorption. J Cell Biol 2007;176(6):877–88.
[13] Riccaboni M, Bianchi I, Petrillo P. Spleen tyrosine kinases: biology, therapeutic targets and drugs. Drug Discov Today 2010;15(13–14):517–30.
[14] Murphy G, Nagase H. Reappraising metalloproteinases in rheumatoid arthritis and osteoarthritis: destruction or repair? Nat Clin Pract Rheumatol 2008;4(3):128–35.
[15] Sondergaard BC, Henriksen K, Wulf H, Oestergaard S, Schurigt U, Brauer R, et al. Relative contribution of matrix metalloprotease and cysteine protease activities to cytokine-stimulated articular cartilage degradation. Osteoarthritis Cartilage 2006;14(8):738–48.
[16] Bay-Jensen AC, Sondergaard BC, Christiansen C, Karsdal MA, Madsen SH, Qvist P. Biochemical markers of joint tissue turnover. Assay Drug Dev Technol 2010;8(1):118–24.
[17] Karsdal MA, Woodworth T, Henriksen K, Maksymowych WP, Genant H, Vergnaud P, et al. Biochemical markers of ongoing joint damage in rheumatoid arthritis - current and future applications, limitations and opportunities. Arthritis Res Ther 2011;13(2):215.
[18] Qvist P, Bay-Jensen AC, Christiansen C, Sondergaard BC, Karsdal MA. Molecular serum and urine marker repertoire supporting clinical research on joint diseases. Best Pract Res Clin Rheumatol 2011;25(6):859–72.
[19] Gelse K, Poschl E, Aigner T. Collagens–structure, function, and biosynthesis. Advanced Drug Delivery Reviews 2003;55(12):1531–46.
[20] Karsdal MA, Schett G, Emery P, Harari O, Byrjalsen I, Kenwright A, et al. IL-6 receptor inhibition positively modulates bone balance in rheumatoid arthritis patients with an inadequate response to anti-tumor necrosis factor therapy: biochemical marker analysis of bone metabolism in the tocilizumab RADIATE study (NCT00106522). Semin Arthritis Rheum 2012;42(2):131–9.
[21] Bay-Jensen AC, Platt A, Byrjalsen I, Vergnoud P, Christiansen C, Karsdal MA. Effect of tocilizumab combined with methotrexate on circulating biomarkers of synovium, cartilage, and bone in the LITHE study. Semin Arthritis Rheum 2014;43(4):470–8.
[22] Leeming D, He Y, Veidal S, Nguyen Q, Larsen D, Koizumi M, et al. A novel marker for assessment of liver matrix remodeling: an enzyme-linked immunosorbent assay (ELISA) detecting a MMP generated type I collagen neo-epitope (C1M). Biomarkers 2011;16(7):616–28.
[23] Barascuk N, Veidal SS, Larsen L, Larsen DV, Larsen MR, Wang J, et al. A novel assay for extracellular matrix remodeling associated with liver fibrosis: an enzyme-linked immunosorbent assay (ELISA) for a MMP-9 proteolytically revealed neo-epitope of type III collagen. Clin Biochem 2010;43(10–11):899–904.

[24] Garnero P, Ferreras M, Karsdal MA, Nicamhlaoibh R, Risteli J, Borel O, et al. The type I collagen fragments ICTP and CTX reveal distinct enzymatic pathways of bone collagen degradation. J Bone Miner Res 2003;18(5):859–67.

[25] Caterson B, Flannery CR, Hughes CE, Little CB. Mechanisms of proteoglycan metabolism that lead to cartilage destruction in the pathogenesis of arthritis. Drugs Today (Barc) 1999;35(4–5):397–402.

[26] Caterson B, Flannery CR, Hughes CE, Little CB. Mechanisms involved in cartilage proteoglycan catabolism. Matrix Biology 2000;19(4):333–44.

[27] Aigner T, Zhu Y, Chansky HH, Matsen III FA, Maloney WJ, Sandell LJ. Reexpression of type II A procollagen by adult articular chondrocytes in osteoarthritic cartilage. Arthritis Rheum 1999;42(7):1443–50.

[28] Bay-Jensen AC, Andersen TL, Charni-Ben TN, Kristensen PW, Kjaersgaard-Andersen P, Sandell L, et al. Biochemical markers of type II collagen breakdown and synthesis are positioned at specific sites in human osteoarthritic knee cartilage. Osteoarthritis Cartilage 2008;16(5):615–23.

[29] Garnero P, Ayral X, Rousseau JC, Christgau S, Sandell LJ, Dougados M, et al. Uncoupling of type II collagen synthesis and degradation predicts progression of joint damage in patients with knee osteoarthritis. Arthritis Rheum 2002;46(10):2613–24.

[30] Oganesian A, Zhu Y, Sandell LJ. Type IIA procollagen amino propeptide is localized in human embryonic tissues. J Histochem Cytochem 1997;45(11):1469–80.

[31] Gudmann NS, Wang J, Hoielt S, Chen P, Siebuhr AS, He Y, et al. Cartilage turnover reflected by metabolic processing of type II collagen: a novel marker of anabolic function in chondrocytes. Int J Mol Sci 2014;15(10):18789–803.

[32] Borth W. Alpha 2-macroglobulin, a multifunctional binding protein with targeting characteristics. FASEB J 1992;6(15):3345–53.

[33] Vassiliadis E, Oliveira CP, Alvares-da-Silva MR, Zhang C, Carrilho FJ, Stefano JT, et al. Circulating levels of citrullinated and MMP-degraded vimentin (VICM) in liver fibrosis related pathology. Am J Trans Res 2012;4(4):403–14.

[34] Bay-Jensen AC, Karsdal MA, Vassiliadis E, Wichuk S, Marcher-Mikkelsen K, Lories R, et al. Circulating citrullinated vimentin fragments reflect disease burden in ankylosing spondylitis and have prognostic capacity for radiographic progression. Arthritis Rheum 2013;65(4):972–80.

[35] Cloos PA, Fledelius C. Collagen fragments in urine derived from bone resorption are highly racemized and isomerized: a biological clock of protein aging with clinical potential. Biochem J 2000;345(Pt 3):473–80.

[36] Cloos PA, Christgau S. Characterization of aged osteocalcin fragments derived from bone resorption. Clin Lab 2004;50(9–10):585–98.

[37] Leeming DJ, Alexandersen P, Karsdal MA, Qvist P, Schaller S, Tanko LB. An update on biomarkers of bone turnover and their utility in biomedical research and clinical practice. Eur J Clin Pharmacol 2006;62(10):781–92.

[38] Veidal SS, Vassiliadis E, Bay-Jensen AC, Tougas G, Vainer B, Karsdal MA. Procollagen type I N-terminal propeptide (PINP) is a marker for fibrogenesis in bile duct ligation-induced fibrosis in rats. Fibrogenesis Tissue Repair 2010;3(1):5.

[39] Rosenquist C, Fledelius C, Christgau S, Pedersen BJ, Bonde M, Qvist P, et al. Serum crosslaps one step ELISA. First application of monoclonal antibodies for measurement in serum of bone-related degradation products from C-terminal telopeptides of type I collagen. Clin Chem 1998;44(11):2281–9.

[40] Huebner JL, Bay-Jensen AC, Huffman KM, He Y, Leeming DJ, McDaniel GE, et al. Alpha C-telopeptide of type I collagen is associated with subchondral bone turnover and predicts progression of joint space narrowing and osteophytes in osteoarthritis. Arthritis Rheumatol 2014;66(9):2440–9.

[41] Leeming DJ, Delling G, Koizumi M, Henriksen K, Karsdal MA, Li B, et al. Alpha CTX as a biomarker of skeletal invasion of breast cancer: immunolocalization and the load dependency of urinary excretion. Cancer Epidemiol Biomarkers Prev 2006;15(7):1392–5.
[42] Bank RA, Robins SP, Wijmenga C, Breslau-Siderius LJ, Bardoel AF, van der Sluijs HA, et al. Defective collagen crosslinking in bone, but not in ligament or cartilage, in Bruck syndrome: indications for a bone-specific telopeptide lysyl hydroxylase on chromosome 17. Proc Natl Acad Sci USA 1999;96(3):1054–8.
[43] Bank RA, TeKoppele JM, Oostingh G, Hazleman BL, Riley GP. Lysylhydroxylation and non-reducible crosslinking of human supraspinatus tendon collagen: changes with age and in chronic rotator cuff tendinitis. Ann Rheum Dis 1999;58(1):35–41.
[44] Mercer DK, Nicol PF, Kimbembe C, Robins SP. Identification, expression, and tissue distribution of the three rat lysyl hydroxylase isoforms. Biochem Biophys Res Commun 2003;307(4):803–9.
[45] Bailey AJ, Peach CM. Isolation and structural identification of a labile intermolecular crosslink in collagen. Biochem Biophys Res Commun 1968;33(5):812–9.
[46] Oxlund H, Barckman M, Ortoft G, Andreassen TT. Reduced concentrations of collagen cross-links are associated with reduced strength of bone. Bone 1995;17(4 Suppl.):365S–71S.
[47] Avery NC, Bailey AJ. Enzymic and non-enzymic cross-linking mechanisms in relation to turnover of collagen: relevance to aging and exercise. Scand J Med Sci Sports 2005;15(4):231–40.
[48] Fledelius C, Johnsen AH, Cloos PA, Bonde M, Qvist P. Characterization of urinary degradation products derived from type I collagen. Identification of a beta-isomerized Asp-Gly sequence within the C-terminal telopeptide (alpha1) region. J Biol Chem 1997;272(15):9755–63.
[49] Mansell JP, Bailey AJ. Abnormal cancellous bone collagen metabolism in osteoarthritis. J Clin Invest 1998;101(8):1596–603.
[50] Sell DR, Monnier VM. Structure elucidation of a senescence cross-link from human extracellular matrix. Implication of pentoses in the aging process. J Biol Chem 1989;264(36):21597–602.
[51] Karsdal MA, Leeming DJ, Dam EB, Henriksen K, Alexandersen P, Pastoureau P, et al. Should subchondral bone turnover be targeted when treating osteoarthritis? Osteoarthritis Cartilage 2008;16(6):638–46.
[52] Karsdal MA, Madsen SH, Christiansen C, Henriksen K, Fosang AJ, Sondergaard BC. Cartilage degradation is fully reversible in the presence of aggrecanase but not matrix metalloproteinase activity. Arthritis Res Ther 2008;10(3):R63.
[53] Everts V, Delaisse JM, Korper W, Jansen DC, Tigchelaar-Gutter W, Saftig P, et al. The bone lining cell: its role in cleaning Howship's lacunae and initiating bone formation. J Bone Miner Res 2002;17(1):77–90.
[54] Reiser K, McCormick RJ, Rucker RB. Enzymatic and nonenzymatic cross-linking of collagen and elastin. FASEB J 1992;6(7):2439–49.
[55] Eyre DR, Paz MA, Gallop PM. Cross-linking in collagen and elastin. Annu Rev Biochem 1984;53:717–48.
[56] Bonde M, Qvist P, Fledelius C, Riis BJ, Christiansen C. Applications of an enzyme immunoassay for a new marker of bone resorption (crosslaps): follow-up on hormone replacement therapy and osteoporosis risk assessment. J Clin Endocrinol Metab 1995;80(3):864–8.
[57] Cloos PA, Christgau S. Non-enzymatic covalent modifications of proteins: mechanisms, physiological consequences and clinical applications. Matrix Biol 2002;21(1):39–52.
[58] Lapolla A, Traldi P, Fedele D. Importance of measuring products of non-enzymatic glycation of proteins. Clin Biochem 2005;38(2):103–15.
[59] Singh R, Barden A, Mori T, Beilin L. Advanced glycation end-products: a review. Diabetologia 2001;44(2):129–46.

[60] Saudek DM, Kay J. Advanced glycation endproducts and osteoarthritis. Curr Rheumatol Rep 2003;5(1):33–40.
[61] Cho SJ, Roman G, Yeboah F, Konishi Y. The road to advanced glycation end products: a mechanistic perspective. Curr Med Chem 2007;14(15):1653–71.
[62] Vander Jagt DL. Methylglyoxal, diabetes mellitus and diabetic complications. Drug Metabol Drug Interact 2008;23(1–2):93–124.
[63] Price CL, Knight SC. Methylglyoxal: possible link between hyperglycaemia and immune suppression? Trends Endocrinol Metab 2009;20(7):312–7.
[64] Turk Z. Glycotoxines, carbonyl stress and relevance to diabetes and its complications. Physiol Res 2010;59(2). 147–5.
[65] Sell DR, Lapolla A, Odetti P, Fogarty J, Monnier VM. Pentosidine formation in skin correlates with severity of complications in individuals with long-standing IDDM. Diabetes 1992;41(10):1286–92.
[66] Viguet-Carrin S, Roux JP, Arlot ME, Merabet Z, Leeming DJ, Byrjalsen I, et al. Contribution of the advanced glycation end product pentosidine and of maturation of type I collagen to compressive biomechanical properties of human lumbar vertebrae. Bone 2006;39(5):1073–9.
[67] Saito M, Marumo K, Kida Y, Ushiku C, Kato S, Takao-Kawabata R, et al. Changes in the contents of enzymatic immature, mature, and non-enzymatic senescent cross-links of collagen after once-weekly treatment with human parathyroid hormone (1–34) for 18 months contribute to improvement of bone strength in ovariectomized monkeys. Osteoporos Int 2011;22(8):2373–83.
[68] Vashishth D. The role of the collagen matrix in skeletal fragility. Curr Osteoporos Rep 2007;5(2):62–6.
[69] Shiraki M, Kuroda T, Tanaka S, Saito M, Fukunaga M, Nakamura T. Nonenzymatic collagen cross-links induced by glycoxidation (pentosidine) predicts vertebral fractures. J Bone Miner Metab 2008;26(1):93–100.
[70] Diab T, Vashishth D. Morphology, localization and accumulation of in vivo microdamage in human cortical bone. Bone 2007;40(3):612–8.
[71] Hein G, Wiegand R, Lehmann G, Stein G, Franke S. Advanced glycation end-products pentosidine and N epsilon-carboxymethyllysine are elevated in serum of patients with osteoporosis. Rheumatology (Oxford) 2003;42(10):1242–6.
[72] Gineyts E, Borel O, Chapurlat R, Garnero P. Quantification of immature and mature collagen crosslinks by liquid chromatography-electrospray ionization mass spectrometry in connective tissues. J Chromatogr B Analyt Technol Biomed Life Sci 2010;878(19):1449–54.
[73] Schwartz AV, Garnero P, Hillier TA, Sellmeyer DE, Strotmeyer ES, Feingold KR, et al. Pentosidine and increased fracture risk in older adults with type 2 diabetes. J Clin Endocrinol Metab 2009;94(7):2380–6.
[74] Yamamoto M, Yamaguchi T, Yamauchi M, Yano S, Sugimoto T. Serum pentosidine levels are positively associated with the presence of vertebral fractures in postmenopausal women with type 2 diabetes. J Clin Endocrinol Metab 2008;93(3):1013–9.
[75] Yamamoto M, Yamaguchi T, Yamauchi M, Sugimoto T. Low serum level of the endogenous secretory receptor for advanced glycation end-products (esRAGE) is a risk factor for prevalent vertebral fractures independent of bone mineral density in patients with type 2 diabetes. Diabetes Care 2009;32(12):2263–8.
[76] Yoshida N, Okumura K, Aso Y. High serum pentosidine concentrations are associated with increased arterial stiffness and thickness in patients with type 2 diabetes. Metabolism 2005;54(3):345–50.

[77] Choi YG, Lim S. Characterization of anti-advanced glycation end product antibodies to non-enzymatically lysine-derived and arginine-derived glycated products. J Immunoassay Immunochem 2009;30(4):386–99.

[78] Taneda S, Monnier VM. ELISA of pentosidine, an advanced glycation end product, in biological specimens. Clin Chem 1994;40(9):1766–73.

[79] Kowluru RA, Atasi L, Ho YS. Role of mitochondrial superoxide dismutase in the development of diabetic retinopathy. Invest Ophthalmol Vis Sci 2006;47(4):1594–9.

[80] Beckman JS, Koppenol WH. Nitric oxide, superoxide, and peroxynitrite: the good, the bad, and ugly. Am J Physiol 1996;271(5 Pt 1):C1424–37.

[81] Beckmann JS, Ye YZ, Anderson PG, Chen J, Accavitti MA, Tarpey MM, et al. Extensive nitration of protein tyrosines in human atherosclerosis detected by immunohistochemistry. Biol Chem Hoppe Seyler 1994;375(2):81–8.

[82] Richardot P, Charni-Ben TN, Toh L, Marotte H, Bay-Jensen AC, Miossec P, et al. Nitrated type III collagen as a biological marker of nitric oxide-mediated synovial tissue metabolism in osteoarthritis. Osteoarthritis Cartilage 2009;17(10):1362–7.

[83] Deberg M, Labasse A, Christgau S, Cloos P, Bang HD, Chapelle JP, et al. New serum biochemical markers (Coll 2-1 and Coll 2-1 NO_2) for studying oxidative-related type II collagen network degradation in patients with osteoarthritis and rheumatoid arthritis. Osteoarthritis Cartilage 2005;13(3):258–65.

[84] Deberg M, Dubuc JE, Labasse A, Sanchez C, Quettier E, Bosseloir A, et al. One-year follow-up of Coll2-1, Coll2-1NO_2 and myeloperoxydase serum levels in osteoarthritis patients after hip or knee replacement. Ann Rheum Dis 2008;67(2):168–74.

[85] Hipkiss AR. Accumulation of altered proteins and ageing: causes and effects. Exp Gerontol 2006;41(5):464–73.

[86] Takigawa M, Okawa T, Pan H, Aoki C, Takahashi K, Zue J, et al. Insulin-like growth factors I and II are autocrine factors in stimulating proteoglycan synthesis, a marker of differentiated chondrocytes, acting through their respective receptors on a clonal human chondrosarcoma-derived chondrocyte cell line, HCS-2/8. Endocrinology 1997;138(10):4390–400.

[87] Fulop Jr T, Larbi A, Dupuis G, Pawelec G. Ageing, autoimmunity and arthritis: perturbations of TCR signal transduction pathways with ageing - a biochemical paradigm for the ageing immune system. Arthritis Res Ther 2003;5(6):290–302.

[88] Wegner N, Lundberg K, Kinloch A, Fisher B, Malmstrom V, Feldmann M, et al. Autoimmunity to specific citrullinated proteins gives the first clues to the etiology of rheumatoid arthritis. Immunol Rev 2010;233(1):34–54.

[89] Mamula MJ, Gee RJ, Elliott JI, Sette A, Southwood S, Jones PJ, et al. Isoaspartyl post-translational modification triggers autoimmune responses to self-proteins. J Biol Chem 1999;274(32):22321–7.

[90] Catterall JB, Barr D, Bolognesi M, Zura RD, Kraus VB. Post-translational aging of proteins in osteoarthritic cartilage and synovial fluid as measured by isomerized aspartate. Arthritis Res Ther 2009;11(2):R55.

[91] Gyorgy B, Toth E, Tarcsa E, Falus A, Buzas EI. Citrullination: a posttranslational modification in health and disease. Int J Biochem Cell Biol 2006;38(10):1662–77.

[92] Anzilotti C, Pratesi F, Tommasi C, Migliorini P. Peptidylarginine deiminase 4 and citrullination in health and disease. Autoimmun Rev 2010;9(3):158–60.

[93] Gudmann NS, Hansen NU, Jensen AC, Karsdal MA, Siebuhr AS. Biological relevance of citrullinations: diagnostic, prognostic and therapeutic options. Autoimmunity 2015;48(2):73–9.

[94] Migliorini P, Pratesi F, Tommasi C, Anzilotti C. The immune response to citrullinated antigens in autoimmune diseases. Autoimmun Rev 2005;4(8):561–4.

[95] Goeb V, Jouen F, Gilbert D, Le LX, Tron F, Vittecoq O. Diagnostic and prognostic usefulness of antibodies to citrullinated peptides. Joint Bone Spine 2009;76(4):343–9.

[96] van Venrooij WJ, Zendman AJ, Pruijn GJ. Autoantibodies to citrullinated antigens in (early) rheumatoid arthritis. Autoimmun Rev 2006;6(1):37–41.

[97] Hansson GK, Libby P. The immune response in atherosclerosis: a double-edged sword. Nat Rev Immunol 2006;6(7):508–19.

[98] Ladenson JH, Tsai LM, Michael JM, Kessler G, Joist JH. Serum versus heparinized plasma for eighteen common chemistry tests: is serum the appropriate specimen? Am J Clin Pathol 1974;62(4):545–52.

[99] Lum G, Gambino SR. A comparison of serum versus heparinized plasma for routine chemistry tests. Am J Clin Pathol 1974;61(1):108–13.

[100] Rossignol P, Cambillau M, Bissery A, Mouradian D, Benetos A, Michel JB, et al. Influence of blood sampling procedure on plasma concentrations of matrix metalloproteinases and their tissue inhibitors. Clin Exp Pharmacol Physiol 2008;35(4):464–9.

[101] Gerlach RF, Uzuelli JA, Souza-Tarla CD, Tanus-Santos JE. Effect of anticoagulants on the determination of plasma matrix metalloproteinase (MMP)-2 and MMP-9 activities. Anal Biochem 2005;344(1):147–9.

[102] Leroy-Matheron C, Gouault-Heilmann M. Influence of conditions of blood sampling on coagulation activation markers (prothrombin fragment 1+2, thrombin-antithrombin complexes and D-dimers) measurements. Thromb Res 1994;74(4):399–407.

[103] d'Audigier C, Delassasseigne C, Robert A, Eschwege V. Underestimation of plasma level of factor V coagulant activity and fibrinogen concentration together with prolonged prothrombin time, activated partial thromboplastin time and thrombin time can result from pre-analytical very low calcium level in citrated sample tube. Int J Lab Hematol 2016;38(1):50–3.

[104] Bonini P, Plebani M, Ceriotti F, Rubboli F. Errors in laboratory medicine. Clin Chem 2002;48(5):691–8.

[105] Papasotiriou M, Genovese F, Klinkhammer BM, Kunter U, Nielsen SH, Karsdal MA, et al. Serum and urine markers of collagen degradation reflect renal fibrosis in experimental kidney diseases. Nephrol Dial Transplant 2015;30(7):1112–21.

[106] Genovese F, Boor P, Papasotiriou M, Leeming DJ, Karsdal MA, Floege J. Turnover of type III collagen reflects disease severity and is associated with progression and microinflammation in patients with IgA nephropathy. Nephrol Dial Transplant 2015.

[107] Thongboonkerd V. Urinary proteomics: towards biomarker discovery, diagnostics and prognostics. Mol Biosyst 2008;4(8):810–5.

[108] Genovese F, Siebuhr AS, Musa K, Gallagher JA, Milan AM, Karsdal MA, et al. Investigating the robustness and diagnostic potential of extracellular matrix remodelling biomarkers in alkaptonuria. JIMD Rep 2015;24:29–37.

[109] Curtis SL, Roberts NB, Ranganath LR. Interferences of homogentisic acid (HGA) on routine clinical chemistry assays in serum and urine and the implications for biochemical monitoring of patients with alkaptonuria. Clin Biochem 2014;47(7–8):640–7.

[110] Petrtýl M, Danešová J, Lísal J, Senolt L, Hulejová H, Polanská M. The initial bearing capacities of subchondral bone replacements considerably contributing to chondrogenesis. Acta Bioeng Biomech 2010;12(3):59–65.

[111] Scott JE, Heatley F. Hyaluronan forms specific stable tertiary structures in aqueous solution: a 13C NMR study. Proc Natl Acad Sci USA 1999;96(9):4850–5.

[112] Saari H, Konttinen YT, Friman C, Sorsa T. Differential effects of reactive oxygen species on native synovial fluid and purified human umbilical cord hyaluronate. Inflammation 1993;17(4):403–15.

[113] Oates KM, Krause WE, Jones RL, Colby RH. Rheopexy of synovial fluid and protein aggregation. J R Soc Interface 2006;3(6):167–74.
[114] Olszta MJ, Cheng X, Jee SS, Kumar R, Kim Y, Kaufman MJ, et al. Bone structure and function: a new perspective. Materials Science and Engineering 2007;05(001):1–40.
[115] Yavorskyy A, Hernandez-Santana A, McCarthy G, McMahon G. Detection of calcium phosphate crystals in the joint fluid of patients with osteoarthritis - analytical approaches and challenges. Analyst 2008;133(3):302–18.
[116] McCarty D. Crystals, joints, and consternation. Ann Rheum Dis 1983;42(3):243–53.
[117] Swan AJ, Heywood BR, Dieppe PA. Extraction of calcium containing crystals from synovial fluids and articular cartilage. J Rheumatol 1992;19(11):1764–73.
[118] Moradi-Bidhendi N, Turner IG. Development of a new technique for the extraction of crustals from synovial fluids. Journal of Materials Science: Materials in Medicine 1995;6(1):51–5.
[119] Larsson S, Lohmander LS, Struglics A. An ARGS-aggrecan assay for analysis in blood and synovial fluid. Osteoarthritis Cartilage 2014;22(2):242–9.
[120] Meyer KC, Raghu G, Baughman RP, Brown KK, Costabel U, du Bois RM, et al. An official American Thoracic Society clinical practice guideline: the clinical utility of bronchoalveolar lavage cellular analysis in interstitial lung disease. Am J Respir Crit Care Med 2012;185(9):1004–14.
[121] Meyer KC. The role of bronchoalveolar lavage in interstitial lung disease. Clin Chest Med 2004;25(4):637–49.
[122] Ryu YJ, Chung MP, Han J, Kim TS, Lee KS, Chun EM, et al. Bronchoalveolar lavage in fibrotic idiopathic interstitial pneumonias. Respir Med 2007;101(3):655–60.
[123] Bronchoalveolar lavage constituents in healthy individuals, idiopathic pulmonary fibrosis, and selected comparison groups. The BAL Cooperative Group Steering Committee. Am Rev Respir Dis 1990;141(5 Pt 2):S169–202.
[124] Noel-Georis I, Bernard A, Falmagne P, Wattiez R. Proteomics as the tool to search for lung disease markers in bronchoalveolar lavage. Dis Markers 2001;17(4):271–84.
[125] Magi B, Bargagli E, Bini L, Rottoli P. Proteome analysis of bronchoalveolar lavage in lung diseases. Proteomics 2006;6(23):6354–69.
[126] Magi B, Bini L, Perari MG, Fossi A, Sanchez JC, Hochstrasser D, et al. Bronchoalveolar lavage fluid protein composition in patients with sarcoidosis and idiopathic pulmonary fibrosis: a two-dimensional electrophoretic study. Electrophoresis 2002;23(19):3434–44.
[127] McCormack FX, King Jr TE, Voelker DR, Robinson PC, Mason RJ. Idiopathic pulmonary fibrosis. Abnormalities in the bronchoalveolar lavage content of surfactant protein A. Am Rev Respir Dis 1991;144(1):160–6.
[128] Cheng G, Ueda T, Numao T, Kuroki Y, Nakajima H, Fukushima Y, et al. Increased levels of surfactant protein A and D in bronchoalveolar lavage fluids in patients with bronchial asthma. Eur Respir J 2000;16(5):831–5.
[129] von Bredow C, Birrer P, Griese M. Surfactant protein A and other bronchoalveolar lavage fluid proteins are altered in cystic fibrosis. Eur Respir J 2001;17(4):716–22.
[130] Lacy P, Lee JL, Vethanayagam D. Sputum analysis in diagnosis and management of obstructive airway diseases. Ther Clin Risk Manag 2005;1(3):169–79.
[131] Paggiaro PL, Chanez P, Holz O, Ind PW, Djukanovic R, Maestrelli P, et al. Sputum induction. Eur Respir J Suppl 2002;37:3s–8s.
[132] Pizzichini E, Pizzichini MM, Efthimiadis A, Evans S, Morris MM, Squillace D, et al. Indices of airway inflammation in induced sputum: reproducibility and validity of cell and fluid-phase measurements. Am J Respir Crit Care Med 1996;154(2 Pt 1):308–17.

[133] Jayaram L, Parameswaran K, Sears MR, Hargreave FE. Induced sputum cell counts: their usefulness in clinical practice. Eur Respir J 2000;16(1):150–8.
[134] Efthimiadis A, Spanevello A, Hamid Q, Kelly MM, Linden M, Louis R, et al. Methods of sputum processing for cell counts, immunocytochemistry and in situ hybridisation. Eur Respir J Suppl 2002;37:19s–23s.
[135] Kidney JC, Wong AG, Efthimiadis A, Morris MM, Sears MR, Dolovich J, et al. Elevated B cells in sputum of asthmatics. Close correlation with eosinophils. Am J Respir Crit Care Med 1996;153(2):540–4.
[136] Ward C, Gardiner PV, Booth H, Walters EH. Intrasubject variability in airway inflammation sampled by bronchoalveolar lavage in stable asthmatics. Eur Respir J 1995;8(11):1866–71.
[137] Ropcke S, Holz O, Lauer G, Muller M, Rittinghausen S, Ernst P, et al. Repeatability of and relationship between potential COPD biomarkers in bronchoalveolar lavage, bronchial biopsies, serum, and induced sputum. PLoS One 2012;7(10):e46207.
[138] Pizzichini E, Pizzichini MM, Kidney JC, Efthimiadis A, Hussack P, Popov T, et al. Induced sputum, bronchoalveolar lavage and blood from mild asthmatics: inflammatory cells, lymphocyte subsets and soluble markers compared. Eur Respir J 1998;11(4):828–34.
[139] Simpson JL, Timmins NL, Fakes K, Talbot PI, Gibson PG. Effect of saliva contamination on induced sputum cell counts, IL-8 and eosinophil cationic protein levels. Eur Respir J 2004;23(5):759–62.
[140] Pizzichini E, Pizzichini MM, Efthimiadis A, Hargreave FE, Dolovich J. Measurement of inflammatory indices in induced sputum: effects of selection of sputum to minimize salivary contamination. Eur Respir J 1996;9(6):1174–80.
[140a] Eastell R, Dijk DJ, Small M, Greenwood A, Sharpe J, Yamada H. Morning vs evening dosing of the cathepsin K inhibitor ONO-5334: effects on bone resorption in postmenopausal women in a randomized, phase 1 trial. Osteoporos Int January 2016;27(1):309–18. http://dx.doi.org/10.1007/s00198-015-3342-4. Epub 2015 Oct 7.
[141] Eastell R. Effects of denosumab on bone turnover markers in postmenopausal osteoporosis. J Bone Miner Res 2011;26:530–7.
[142] Engelke K, Fuerst T, Dardzinski B, Kornak J, Ather S, Genant HK, et al. Odanacatib treatment affects trabecular and cortical bone in the femur of postmenopausal women: results of a two-year placebo-controlled trial. J Bone Miner Res 2015;30(1):30–8.
[143] Rosen CJ. Treatment with once-weekly alendronate 70 mg compared with once-weekly risedronate 35 mg in women with postmenopausal osteoporosis: a randomized double-blind study. J Bone Miner Res 2005;20:141–51.
[144] Black DM. One year of alendronate after one year of parathyroid hormone (1–84) for osteoporosis. N Engl J Med 2005;353:555–65.
[145] Fleischmann RM, Halland AM, Brzosko M, Burgos-Vargas R, Mela C, Vernon E, et al. Tocilizumab inhibits structural joint damage and improves physical function in patients with rheumatoid arthritis and inadequate responses to methotrexate: LITHE study 2-year results. J Rheumatol 2013;40(2):113–26.
[146] Siebuhr AS, Bay-Jensen AC, Leeming DJ, Plat A, Byrjalsen I, Christiansen C, et al. Serological identification of fast progressors of structural damage with rheumatoid arthritis. Arthritis Res Ther 2013;15(4):R86.
[147] Jubb RW, Fell HB. The breakdown of collagen by chondrocytes. J Pathol 1980;130(3):159–67.
[148] Rousseau JC, Garnero P. Biological markers in osteoarthritis. Bone 2012;51(2):265–77.
[149] Christensen AF, Horslev-Petersen K, Christgau S, Lindegaard HM, Lottenburger T, Junker K, et al. Uncoupling of collagen II metabolism in newly diagnosed, untreated rheumatoid arthritis is linked to inflammation and antibodies against cyclic citrullinated peptides. J Rheumatol 2010;37(6):1113–20.

[150] Karsdal MA, Byrjalsen I, Henriksen K, Riis BJ, Christiansen C. A pharmacokinetic and pharmacodynamic comparison of synthetic and recombinant oral salmon calcitonin. J Clin Pharmacol 2009;49(2):229–34.
[151] Garnero P, Landewé R, Boers M, Verhoeven A, van der Linden S, Christgau S, et al. Association of baseline levels of markers of bone and cartilage degradation with long-term progression of joint damage in patients with early rheumatoid arthritis: the COBRA study. Arthritis Rheum 2002;46(11):2847–56.
[152] Landewe RB, Boers M, Verhoeven AC, Westhovens R, van de Laar MA, Markusse HM, et al. COBRA combination therapy in patients with early rheumatoid arthritis: long-term structural benefits of a brief intervention. Arthritis & Rheumatism 2002;46(2):347–56.
[153] Downs JT, Lane CL, Nestor NB, McLellan TJ, Kelly MA, Karam GA, et al. Analysis of collagenase-cleavage of type II collagen using a neoepitope ELISA. J Immunol Methods 2001;247(1–2):25–34.
[154] Little CB, Hughes CE, Curtis CL, Janusz MJ, Bohne R, Wang-Weigand S, et al. Matrix metalloproteinases are involved in C-terminal and interglobular domain processing of cartilage aggrecan in late stage cartilage degradation. Matrix Biol 2002;21(3):271–88.
[155] Otterness IG, Swindell AC, Zimmerer RO, Poole AR, Ionescu M, Weiner E. An analysis of 14 molecular markers for monitoring osteoarthritis: segregation of the markers into clusters and distinguishing osteoarthritis at baseline. Osteoarthritis Cartilage 2000;8(3):180–5.
[156] Hellio Le Graverand MP, Brandt KD, Mazzuca SA, Katz BP, Buck R, Lane KA, et al. Association between concentrations of urinary type II collagen neoepitope (uTIINE) and joint space narrowing in patients with knee osteoarthritis. Osteoarthritis Cartilage 2006;14(11):1189–95.
[157] Briot K, Gossec L, Kolta S, Dougados M, Roux C. Prospective assessment of body weight, body composition, and bone density changes in patients with spondyloarthropathy receiving anti-tumor necrosis factor-alpha treatment. J Rheumatol 2008;35(5):855–61.
[158] Briot K, Roux C, Gossec L, Charni N, Kolta S, Dougados M, et al. Effects of etanercept on serum biochemical markers of cartilage metabolism in patients with spondyloarthropathy. J Rheumatol 2008;35(2):310–4.
[159] Hu XR, Cui XN, Hu QT, Chen J. Value of MR diffusion imaging in hepatic fibrosis and its correlations with serum indices. World J Gastroenterol 2014;20(24):7964–70.
[160] Schuppan D. Structure of the extracellular matrix in normal and fibrotic liver: collagens and glycoproteins. Semin Liver Dis 1990;10(1):1–10.
[161] Gressner AM, Weiskirchen R. Modern pathogenetic concepts of liver fibrosis suggest stellate cells and TGF-beta as major players and therapeutic targets. J Cell Mol Med 2006;10(1):76–99.
[162] Schuppan D, Ruehl M, Somasundaram R, Hahn EG. Matrix as a modulator of hepatic fibrogenesis. Semin Liver Dis 2001;21(3):351–72.
[163] Cassiman D, Roskams T. Beauty is in the eye of the beholder: emerging concepts and pitfalls in hepatic stellate cell research. J Hepatol 2002;37(4):527–35.
[164] Friedman SL. Seminars in medicine of the Beth Israel Hospital, Boston. The cellular basis of hepatic fibrosis. Mechanisms and treatment strategies. N Engl J Med 1993;328(25):1828–35.
[165] Libbrecht L, Cassiman D, Desmet V, Roskams T. The correlation between portal myofibroblasts and development of intrahepatic bile ducts and arterial branches in human liver. Liver 2002;22(3):252–8.
[166] bdel-Aziz G, Rescan PY, Clement B, Lebeau G, Rissel M, Grimaud JA, et al. Cellular sources of matrix proteins in experimentally induced cholestatic rat liver. J Pathol 1991;164(2):167–74.
[167] Gressner AM. Perisinusoidal lipocytes and fibrogenesis. Gut 1994;35(10):1331–3.
[168] Nielsen MJ, Veidal SS, Karsdal MA, Orsnes-Leeming DJ, Vainer B, Gardner SD, et al. Plasma Pro-C3 (N-terminal type III collagen propeptide) predicts fibrosis progression in patients with chronic hepatitis C. Liver Int 2015;35(2):429–37.

Index

'*Note:* Page numbers followed by "f" indicate figures and "t" indicate tables.'

V

Printed in the United States
By Bookmasters